Donahue, Brian
Reclaiming the commons

Yale Agrarian Studies

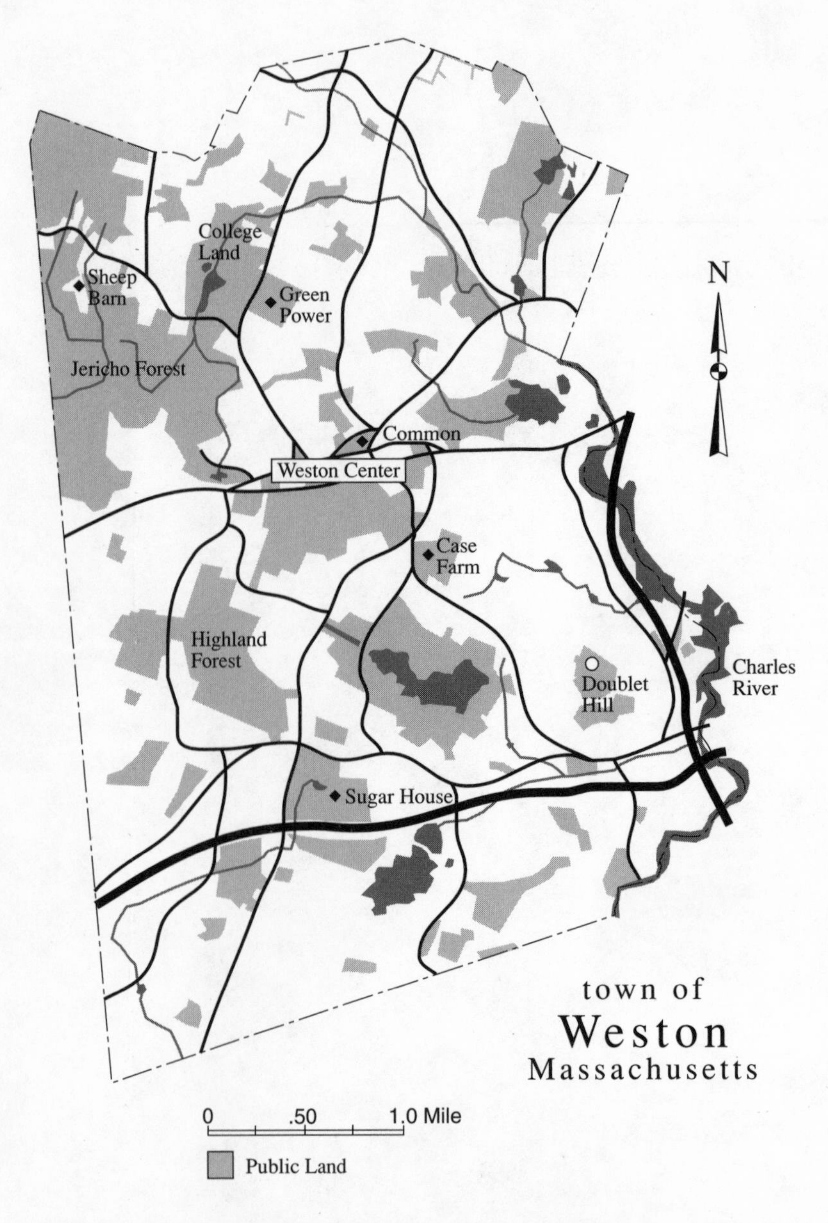

Publicly owned land in Weston, Massachusetts. (Adapted by Cathleen Daley from the Weston Forest and Trail Association Map of Parks, Conservation, and Municipal Land in the town of Weston, 1993)

Reclaiming the Commons

Community Farms & Forests in a New England Town

Brian Donahue

Foreword by Wes Jackson

Yale University Press

New Haven & London

Published with assistance from the foundation established in memory of Philip Hamilton McMillan of the Class of 1894, Yale College.

Designed by Rebecca Gibb.
Set in Adobe Caslon type by à la page, New Haven, Connecticut.
Printed in the United States of America.

Library of Congress Cataloging-in-Publication Data
Donahue, Brian, 1955–
Reclaiming the commons : community farms and forests in a New England town / Brian Donahue : foreword by Wes Jackson.
p. cm. — (Yale agrarian studies)
Includes bibliographical references (p.) and index.
ISBN 0-300-07673-8 (alk. paper)
1. Commons—Massachusetts—Weston. 2. Collective farms—Massachusetts—Weston. 3. Community forests—Massachusetts—Weston. 4. Land use, Rural—Environmental aspects—Massachusetts—Weston. I. Title. II. Series.
HD1289.U6D66 1999
333.2—dc21
98-49122
CIP

A catalogue record for this book is available from the British Library.

The paper in this book meets the guidelines for permanence and durability of the Committee on Production Guidelines for Book Longevity of the Council on Library Resources.

10 9 8 7 6 5 4 3 2 1

For Faith

Contents

Foreword

Wes Jackson

Smart children of parents who have spent much of their productive lives worrying about (and maybe now and then working on) the global problems of population growth, resource depletion, and pollution of nature's sinks and sources have been known to ask questions similar to the following: "And so, Dr. Doom, what can *we* do about it where *we* live?" How *can* a suburban Sierra Club member who frets over spotted owls and pesticides act beyond writing his or her congressperson about the owls and avoiding the use of pesticides on the lawn and garden? I have always supposed that a primary source of authenticity for proposed solutions to these problems must come from those who have had a serious engagement with the land itself. It is this historical engagement that generates the evolving cultural arrangements for the harvest of the stored energy from that great centralized fusion reactor in the sky 93 million miles away. And so, here at last is a book dedicated to getting people in the suburbs to engage with their land.

Brian Donahue walked onto the grounds of the Land Institute in January 1994 to become its education director with a promise to stay three years. He did, and while he was there, he changed minds. But then he returned to the same Boston suburb from which he had come, leaving many of us at the institute to wonder why such a seemingly sane environmentalist would want to do that once he had tasted the Kansas prairie, minimal traffic, and far superior weather. (Why indeed?) Well, this book has supplied a rather complete answer to our questions about his sanity. We shouldn't have been surprised, for his depth as an environmental historian constantly generated useful insights and even categorical surprises that made us all aware of the importance of his chosen field to an environmental curriculum. The following pages are vintage Brian Donahue, bringing social history and natural history together, subjects usually treated separately. Donahue's treatment of Weston, Massachusetts, from precolonial times to the present represents a prototype of the sort of scholarship and action-on-scholarship that America needs. It is about suburbia, that modern Leviathan with an irrational settlement pattern, prime-farmland-gobbling, mall-producing, loneliness-generating, xenophobic, consumptive life where secular materialism has become the national religion. The majority of Americans are either already in suburbia or headed there. It is the world that defines our being, the world that even the nonsuburbanite is pulled into. It is the world that investors in economic growth gain from, a gain that dooms our culture and forces our descendants to pay the bills in the long run.

Rural landscapes and their declining small towns—and for that matter burned-out inner cities—present problems obvious to any who care to look, even while driving past over the speed limit. In the country, soil erodes and outbuildings collapse while the farmyard is increasingly adorned with huge tractors, disks, harvesters, and industrial machinery built to lay on the ammonia fertilizer and synthetic pesticides. The problem is greater than the fact that so many major U.S. cities count large populations of the poor and drug-addicted. These are just two of the many symptoms of our cultural sickness, which includes poor land use and other easily recognized ecological sicknesses in need of healing.

The suburbs, on the other hand, are "where it's at." In suburbia, children of the affluent are raised with the idea that they are in the midst of a legitimate American dream. In Donahue's Weston, most of the kids are fit, play soccer, excel in a first-rate school system, and perhaps after a minor bout of adolescent alienation go on to live affluent lives, fretting a little about the loss of rainforest. At the other extreme are suburban kids dreadfully overweight and more of the TV-watching, arcade-playing type, attending schools that reward minimal compliance. Whatever the situation, most failures on the suburban landscape are due to economic success. The suburban dream rests on the social and environmental nightmares that haunt the inner cities and the harrowed countryside.

Weston, Massachusetts, is a suburb. Why would an environmentalist begin with a place like Weston, where the inhabitants perceive that life is as good as it gets? For that very reason Donahue did become interested; he saw that in suburbia people are in the "belly of the beast." What suburbia needs does not lend itself to bumper sticker or T-shirt environmentalism. Sporting an artfully silk-screened "Think Globally—Act Locally" T-shirt and then limiting that action to recycling plastic in six categories won't do it. The suburbs exist because there have been and still are industrial vandals in our land, people we never seem to nab precisely because we are addicted to what they have to sell or offer. And so we continue to buy stickers and T-shirts with clever, catchy phrases and unwittingly help to fuel industrialization's movement of people—including rural people—to the suburbs.

The industrial mind has promoted the idea that a sufficiency of capital should take precedence over a sufficiency of people. But what about the idea that most people *need* a landscape, and what about the fact that essentially all agricultural landscapes in the United States *need* more people? What about the fact that most kids want to work in a meaningful way but cannot because of our settlement patterns? Work is one of the most natural things to do if it is of the right kind. On a recent hot June day my three-year-old grandson outlasted me at picking in the cherry tree and later in the raspberries as well. At the end of his two-day visit he rewarded my effort by declaring me his best buddy. It made me feel good even though I know that anyone else in my place would have been declared his best buddy. The sandpile and his toys were far less interesting to him those two days than what amounted to a harvest of contemporary sunlight and resultant health. Our evolution shaped that predisposition in him, and the context of those June days allowed its playing out.

Nearly twenty years ago Mary Catherine Bateson blurted out in a conversation of which I was a part, "Boredom has to be taught." It was a stunning and, I believe, accurate statement. Until that time I thought boredom was a natural derivative of not having something to do or having work that can go on too long. And of course it is, but clearly there is more to it than that. We teach boredom to our suburban children in countless ways. Just look at all the schools without windows or the frequency with which teachers tell kids to pay attention and quit looking *out* the windows. The areas in which we overstimulate are seldom berry- or cherry-picking. Children need reality, not virtual reality, and there are plenty of examples you can read about in these pages. Children need to understand the nature of source, whether it is wood for a stove or food for their bellies. Donahue's nonreductive way of showing connections to the young can be replicated; indeed, it has the potential to relegate the environmental bumper sticker and T-shirt to some museum of the future dedicated to the period of the environmentally and socially naïve.

If America doesn't save agriculture with a sufficiency of people, wilderness is doomed. It is doomed as an artifact of civilization because hungry, formerly civilized people will encroach upon it and destroy it to meet immediate needs. What better place to teach the essentials of agriculture and forestry than in the suburbs, where most people are located?

Knowing the history of a piece of land from presettlement times to the present, the environmental historian has nevertheless been limited in role. Missing has been the more prescriptive work that would shape people's actions in accord with the limits and possibilities of the particular lands they occupy. But now Brian Donahue has created a new genre, a success story out of the suburbs sure to give heart to sociologist and ecologist alike. Here it is, an intellectual's and an activist's approach.

Preface

I am an environmentalist. Like many environmentalists, I live in the suburbs—and don't feel good about it. Candid suburban environmentalists recognize that our way of life epitomizes industrial society's reckless consumption of nature. It is not good that we live in one place and work in another, and daily pollute both (and all those places in between, and the world in general) as we commute. It is not good that we live in such a sprawling fashion. It does not feel good to demand such a rich diet of resources—but what can we do about it?

Of course our cities could be made economically healthy, livable, and green if more of us bent our energies to revitalizing them rather than fleeing from them. From time to time we see movement in that direction in certain city neighborhoods—although it often seems to bypass or evict those who need it most. In a better world, those who prefer urban life would live closer to where they work. Suburbanization driven by the dereliction of the cities is a double crime.

But many of us are genuinely drawn to the countryside. We feel we are agrarians at heart, but because we are not farmers we end up in the suburbs. We carry the suburbs with us wherever we go. It is apparent that much of our economy could be decentralized, allowing many more people who are not farmers and loggers to live in rural areas. Indeed, this transformation is well under way. But it is not going well for the land. Just as the electronic revolution was supposed to create the paperless office and instead heralded a blizzard of paper, telecommuting, in today's cheap energy economy, spawns ever more automotive traffic. People drive farther and farther from their dream homes to consume the goods and services they crave; one spouse telecommutes but the other drives seventy miles to town. Scattered houses chop up the land, subdividing farmsteads and fragmenting forests. Before us rolls a tide of suburbanization that dwarfs earlier outward ripples, engulfing the countryside as far as the eye can see.

This book urges that something be done about the suburbs, and that we suburbanites do it. We must learn how to protect (from ourselves) those things about the countryside that we most admire. We must learn how to be environmentally responsible citizens. But this is not a book about the 101 easy changes that won't cramp our modern lifestyles. Still less is it about renouncing modern life and living "off-the-grid," on self-sufficient farms—not that I have anything against those who do such worthy things. This is not a book about taking individual action to salve a stricken conscience. This book is about improving the suburbs as *communities* of people and land. It is about common action.

Reclaiming the Commons argues two things. First, suburbanites must work to protect forest and farmland as they resettle the countryside. Not just some minor open space: a major part of the forest, prairie—whatever the native landform may be—should be acquired as local commons. Farmland that is subject to development pressure should be protected by easements. Suburbanites should curb their craving for large private estates and concentrate instead upon securing a healthy and attractive community landscape within which to live.

Second, this land should be protected not simply for passive ecological, educational, or recreational purposes. It should also be sustainably used for productive purposes. We hear a lot about the need for a sense of place, about becoming more familiar with the ecosystems that surround us and with the history of our communities. I am all for that, but it is nowhere near enough. We need *engagement* with place.

Reclaiming the Commons calls for the protection of several thousand acres of forest and farmland in every rural community. It then urges the establishment of one nonprofit *community farm* in each of these places. The purpose of community farm and forest programs is to involve as many people as possible with caring for the land around them.

I need to state here (as I will again) that I am *not* calling for the forced expropriation of private land. Neither am I claiming that collective farms are superior to private farms. On the contrary, I think small private farms are the best thing in the world, and I wish to see far more of them. Suburbanites presently destroy farms as they arrive in rural towns. Instead, they should help shoulder the burden of protecting private farms, by funding the purchase of easements by the community. They should also help purchase common forests, to be controlled by the local community but logged primarily by private contractors. I am calling for a better balance between private and common landholding, and I am calling for *one* community farm among the many private enterprises in each place.

I make these proposals based on practical experience. For most of the past quarter century I have lived in Weston, Massachusetts, a suburb of Boston. With my friends and neighbors I have been involved in protecting some two thousand acres of land and particularly in organizing and running two community farms, Green Power and Land's Sake, that employ young people to care for that land. At present these farms, now merged, cultivate twenty-five acres of fruits, flowers, and vegetables by organic methods; harvest firewood and timber within some fifteen hundred acres of town forest; maintain sixty-five miles of trails; make maple syrup and apple cider; and run a variety of educational programs. In the past we have kept a flock of sheep, worked with draft horses, cultivated an orchard, run a sawmill—and we hope to return to these and many other projects in the future, as we continue to find what works best in our situation. Our program has grown by trial and error, with plenty of error, as you shall read. I cannot think of a crucial agricultural or political blunder that I have not made, with the single exception of using white birch for fenceposts. But I am still young. I expect to commit many more, of which I am now blissfully unaware.

This book describes that journey from apprenticeship to competence, if not complete mastery. I will recount our adventures in market gardening, livestock husbandry, syruping and cidering, and sustainable forestry, in a conceptual order running from the most to the least intensive use of land, from the garden through the back pasture to the woods. Each chapter arrives at a conundrum that had to be overcome, a fallen tree blocking the trail. At that point, I pause to examine Weston's history, in order to explore the long-standing challenges facing sustainable engagement with the land in this corner of the world. As a result, embedded in *Reclaiming the Commons* is a compact ecological history of the region, from the departure of the last glacier to the truck farms that surrounded Boston well into the present century. Or rather, from the truck farms back to the glacier, because I have laid the book out so that I have to dig into this history backwards, perhaps to enliven my own comfortable pedantry. These historical sections are based on research I have been conducting for as long as I have been farming and also expect never to complete.

Each chapter concludes with a discussion of the established project as it reflects Land's Sake's four guiding principles: ecology, economics, education, and esthetics. Ecology addresses the issue of sustainable use of land, in the very broadest (and hence most exacting) sense of that term. Economics examines how each project endeavors to pay for itself, as far as is possible given the present market. Education relates how people, particularly young people, become involved with the land. Esthetics explores the small ways in which engaging with the land satisfies the soul, even in the midst of the suburbs. How we do what we do emerges from the tensions among these imperatives.

The story is laid out neither in the neat chronological order in which a historian might tell it, nor in the nicely rounded seasonal way a natural history writer might tell it. Rather, it grows organically, somewhat as one might delve into it over the years. Call it a farmer's approach. This is in fact the way I have reached an imperfect understanding of the place where I live and work.

The book concludes with a call for all places to protect common land and establish community farms. It calls for a new *common agrarianism* within suburbanizing places. The suburbs are where most Americans live, and where we environmentalists will make perhaps our most lasting mark on the world. If we have a home landscape, this is it.

Obviously, a book that represents more than two decades of work both in Weston and within a wider community of environmentalists and agrarians leaves me with many more people to thank than I possibly can here. Within Weston I would like to acknowledge William Elliston, Julie Hyde, and George Bates, as three worthy representatives of half a century of dedicated effort to protect and care for conservation land in our town. I would like to thank Bill McElwain, Doug Henderson, Martha Gogel, and Bear Burnes for years of selfless determination to make Green Power and Land's Sake succeed. Thanks to Jerry Howard, Kippy Goldfarb, Neil Baumgarten, Susan Campbell, Tom Gumbart, Cathy Daley, and the Weston Forest and Trail Association for providing illustrations. Thanks to Bill McElwain, Doug Henderson, Julie Hyde, Ed and Polly Dickson, and Carl Johnson for discussing the history of Weston, Green Power, and Land's Sake with me. Thanks to my editors, Jean Thomson Black and Lawrence Kenney, for their good work. Finally, I would like to thank Bill Vitek, Wes Jackson, Don Worster, Wendell Berry, Molly Anderson, Doug Henderson, Bill McElwain, Julie Hyde, John Potter, Faith Rand, and my parents, Thomas and Esther Donahue, for commenting on the manuscript. I would venture to say that few aspiring authors have been blessed with such a splendid crew of critics. In spite of their best efforts, I am confident that my literary faults remain as comprehensive as my agricultural ones.

This book is also a love story. It is not only for Faith, but from her as well.

Reclaiming the Commons

Introduction

Wilderness and Suburbia

I am a child of the suburbs. I grew up north of Pittsburgh, Pennsylvania, in the 1960s on what was then the outer edge of postwar suburban expansion. Within the horizons of childhood a few years are an eternity, so I was only dimly aware that I was living on the *cutting* edge of suburbia and that the rural world had gone to seed and was being mowed down all around me. As I remember it, much of my youth took place outdoors, unsupervised. My boyhood was spent roaming free and exploring the woods and ragged pastures beyond my backyard; I built dams and hideouts, picked wild berries, and caught turtles, toads, snakes, and even one inept chipmunk. True, I went to school in a new consolidated suburban school system that could scarcely build schools fast enough to keep up with the baby boom. I played Little League baseball, went trick-or-treating through endless tract developments swarming with my fellow boomers, and did all the things that mark a stereotypical suburban upbringing. But I also enjoyed the unfettered access to nature that was part of America's earlier, idyllic agrarian past. I am therefore also a child of the rural American countryside, but not of a living rural culture. When I was grown, I discovered I was an agrarian orphan.

Very little active farming was going on anymore where I grew up. I did not grasp it at the time, of course, but the place I loved as a boy was in reality an abandoned, dying agricultural landscape. Western Pennsylvania is beautifully hilly, mostly too hilly for modern commercial farming. Sometime in the decades before I was born, farming largely ceased in my native place. The narrow creek bottoms and steep valley sides were heavily wooded, while the uplands were mostly farmed out and growing up in brush. Only a few of the wider bottoms and gentler slopes were still cultivated or at least periodically mowed. Here and there were old orchards almost completely overgrown with young maple and ash, the derelict fruit trees still bearing a few tart, wormy apples that my friends and I hurled at each other after one bite. Most of the woods and fields where I played were simply waiting to be developed. By now, I suppose most of them have been.

My parents did not come from the region, so I grew up without local roots. I think this, above all else, defines suburban existence everywhere: the particular place makes no difference. Suburbia is the condition of residing outside the city proper with little functional connection to one's neighbors, aside from the schools, and almost no functional connection to the land, aside from an optional backyard garden. Once upon a time suburban families at least mowed their lawns, trimmed their shrubbery, and planted young trees around their new houses; today even that work is largely hired out to professionals. The residents of these nonplaces typically move once every decade, if not twice. Such transient people tend to form only temporary ties to neighbors and nature, cultural and ecological slipknots. I grew up with no sense of the history of the land where I lived or of the people who had lived there for generations, whose German and Scotch-Irish ancestors had settled the region in the eighteenth century. I was exposed early to nature, to gardening and birdwatching, by my parents. I was taken on walks in the woods and encouraged to splash in the streams while still a toddler, for which I am grateful. But however much I learned to love the face of the land, I did not identify with any culture that was bred in the body of the land. And indeed, very little remained to identify with.

the liberal politics of my parents
ronmental movements of the late
nths shy of fifteen and finishing my
the first Earth Day teach-ins took
my high school years developing a
th my friends I formed an ecology
cling programs, litter and stream
to combat air and water pollution.
hool I felt I was beyond all that. My
g industrial society and toward the
ecological exploration of the outly-
inded friends. We spent our week-
t a cabin on an Audubon sanctuary
eral of us returned to work as sum-
. We took long camping trips into
he central part of the state, walking
road as you can get, east of the Mis-
f outward rebellion against our par-
andoned the American dream. We
would allow us to surpass the affluent suburban lifestyle in which we had been raised—quite the contrary, we wanted to live more simply, immersed in nature. In one way or another, many of us have.

Like many environmentalists at the time, we believed that industrial society had upset the balance of nature so severely that it was fast approaching a crisis. We thought our way of life was going to have to be radically simplified by about 1980; certainly by the year 2000. We assumed we were going to live through one of the great transformations in human history. A sweeping postindustrial revolution was coming. I welcomed this prospect, although enough raw twentieth-century history had been drilled into me by my father to make clear that if this happened it might well be a time of nightmarish social chaos, one in which our kind would be among the first to be taken out and shot. But I vaguely planned to get away from all that and just watch it come down.

When I envisioned an ideal place to live, it was on the edge of the wilderness—perhaps in Canada or out west or even in a small hollow in the mountains of Pennsylvania. Someplace with plenty of trees and plenty of snow and not too many people around. It was a Thoreauvian ideal: to live simply on a small patch of cultivated land, practically invisible, with undisturbed wilderness at my back. My friend Fred Huemmrich and I called this the Cabin in the Woods Fantasy and wrote each other letters all through college detailing how we might bring it about. Satisfaction and meaning in life were to be found as much as possible through direct, personal contact with the natural world, as unobtrusive observers. This, we believed, is how a much-chastened human race would eventually have to live, or better yet would *want* to live. I searched for that connection myself, and quite often, while walking through the woods, especially in rain or snow, I seemed to find it. The city and suburbs were far from my mind. They were dinosaurs, doomed to extinction. In my imagination, I was outward bound for the big sticks.

In the summer of 1975, I worked as a motel maid in Aspen, Colorado. I had vacationed in the Rocky Mountains a few times with my family and was strongly drawn to return to them on my own. I recalled the sharp smell of conifers and the taste of cold, clean water at high altitude. Cleaning the rooms was not too taxing and was over by midafternoon, which left the maids plenty of time for reading, writing, cooking, playing music in our basement pad—and taking walks. On my day off I took all-day hikes into the mountains. I was a tireless walker, and I got about as far away from civilization as it is possible to get in a single day. I carried a small cotton haversack with my binoculars, a map and bird book, a sweater, and a few pieces of fruit.

That summer I explored the valley of Hunter Creek to the northeast of Aspen. Usually I walked alone. One afternoon late in June I stood in a high park north of the valley, looking east toward the headwaters. Across a great bowl where Midway and Hunter creeks ran together reared up the Williams Mountains, a craggy, convoluted range of stony peaks thirteen thousand feet high. Several thunderstorms at once swept past their crests and down the valleys. There it was: Shangri-la, completely wild and apparently all but unknown. I scarcely heard anyone in Aspen talk about this range: for some reason the Williams had not become a popular wilderness destination. Not as spectacular, perhaps, as some of the snow-capped fourteen-thousand-foot peaks that lay in other directions. What could be better? An unheralded, undesignated, unappreciated, unknown wilderness all my own. A true wilderness, in other words.

In early August, my fellow-maid Chuck Holton and I quit our jobs at the motel and spent a few days camping in the Williams Mountains before heading back east. We informed the Forest Service of our plans, threw our sleeping bags and pads, some freeze-dried trail food, a cooking kit, and a polytarp into a couple of old canvas packs and stomped up there. We made camp beside a tributary of Midway Creek at treeline, about 11,500 feet, just below the peaks. The night was clear, and the heavens were spectacular—fuller of stars than I ever imagined. The next day Chuck and I had the Williams Mountains to ourselves. We climbed several of the peaks and then dropped into the long, curving valley of Hunter Creek on the far side, following it down for several miles, right from its source. Late in the afternoon we climbed out of the steep valley and over the mountains again, descending to our camp at sunset. It was a long, grueling walk, with a tortuous climb near the end. Chuck had not done as much walking in the high country as I had that summer and was not used to the altitude. I remember him doggedly forcing himself up the steep slope half a step at a time with the help of his walking stick, planting it squarely in front of his body and hauling himself up another notch with both hands, his empty gaze turned inward until he reached the pass.

But while my body was better adapted to the mountain air, something had gone inexplicably wrong with my mind. Toward the end of that walk I became acutely depressed. It was as if the physical scale of what I was traversing demanded that I cover the same amplitude of emotion in a single day. Having been up, I got about as down as I have ever been, and what made it worse was that it seemed completely at odds with how I should have felt. I was on the most awe-inspiring wilderness trek of my life, the culmination of a fine summer of mountain exploration. I was exactly where I wanted to be, doing exactly what I most wanted to do. I was not in the midst of any personal crisis. Yet all at once I wanted desperately to be gone from there, to be home, wherever that was—in Michigan where my parents lived, or in

Pennsylvania where I had grown up, or even back in Boston getting ready for another year at Brandeis. But time slowed nearly to a stop and stranded me right there in the interminable present. The moment seemed completely void of meaning, with no hope of escape. The wait for life to begin again was too long to bear. It took all the resources I could muster to walk down the last slope to camp, to go through the motions of cooking and eating, of simply breathing. Where in the world was this coming from?

Looking back, I can suggest some likely explanations of this odd episode. I have heard that high altitude can bring on depression, and that may have been partly to blame. Because we were about to leave, maybe I was already feeling partly absent. It is possible to overdose on wilderness and to be overwhelmed by too much scenery on a grand scale, and that was how I diagnosed my complaint at the time. The next day we decided not to take another long hike as we had planned, but to just hang out. Chuck stretched himself out beside the stream in his hammock, finishing Rollo May's *Love and Will* (one of the thickest books I have ever seen). I enjoyed the image of Chuck reading a hardbound library book up there in the wilderness, but I was in no mood to read myself. I climbed up the steep, rocky ledges of the mountainside a few hundred feet above camp and sat down on a narrow shelf to think things over. There in the thin soil were strange, beautiful alpine wildflowers that I had never seen before and, since I had no field guide, could not identify. Sweeping vistas of nature need to be balanced with immersion in the intimate details of nature, and those wildflowers, whose names I never knew, seemed to help.

I was no longer so depressed, but I was still lonely and perplexed. I couldn't figure out why I felt so unsatisfied in the midst of the wilderness I loved. I sat looking from my eyrie over the valley of Midway Creek and leagues of untroubled forest, my back to good mountain stone, pondering my condition. Something had been nagging at me as I sat on similar overlooks during my walks that summer, and now I examined it more closely. My infatuation with wilderness was misguided. The trees were unmoved by my presence and unresponsive. I said I loved them, but they did not love me. What did my so-called love for them mean, in practical terms? There was no actual possibility of a direct emotional connection between myself and wilderness.

That was my problem: I was walking in a magnificent world of unspoiled nature, but I had no meaningful connection to it. Wallace Stegner has written that civilized people go to the wilderness not to find themselves but to disappear; and in retrospect I think that was exactly what was bothering me. I was lonely because I did not belong there. It was partly the old paradox of being a human being in a wilderness, a place defined by the absence of human beings, and hence contradicting it by my very presence. But it went beyond that. This was the matter: there was a terrible problem between my species and nature, but the answer to it could not be found out there, where people were absent.

Wilderness is marginal to solving the ecological dilemma of humanity. That was the hard truth. This is not to say that wilderness itself is bad, but rather that by itself wilderness offers no useful counterbalance to industrialism, and industrialism offers no sanctuary to it. The thirst for Henry Thoreau's "tonic of wildness," such historians as Roderick Nash and Samuel Hayes have told us, has grown in the past century in response to the unprecedented rise of an urban, industrial civilization. The wilderness has become a place for the spiritual solace and regeneration of the overcivilized. The ability to appreciate wilderness, we are told, is a luxury belonging to those who need no longer struggle with nature everyday. That is, the condition for our love of nature is our removal from direct contact with nature, which was really pretty unpleasant, we are assured. Disease, hunger, violence, poverty, dirt: that is the true state of nature. This is the message of those who are comfortable with the idea that modern civilization and wilderness can coexist and that wilderness in fact depends on civilization for its definition.[1]

Are only those who are the complete masters of nature really free to love it, to safely admire and revel in its wildness? This was the half-truth in which I was entangled with my passion for wilderness. I didn't believe it then, and I don't believe it now—but I saw I was doing nothing to disprove it, either. It was an unacceptable idea to me because I was then and am now convinced that ultimately urban, industrial civilization as we have constructed it is incompatible with wilderness, with a healthy natural world in general. If left unchecked, the economic and technological forces that have driven the rapid rise of industrial society will directly consume all of nature that can be brought to market and indirectly degrade the rest. Retreating to the shrinking wilderness, I saw, was a completely inadequate response to this problem. Moreover (as thoughtful wilderness advocates agree), striving to protect wilderness was a hopeless task, unless the root problem was addressed. The problem, and its possible solutions, lay beyond the wild mountain ranges, down in the flatlands where people went about their business; lay in mending the ways we daily gained our living from nature. What we needed was not to protect whatever amount of wilderness: our culture was incapable of living in anything approximating wilderness again. What we needed was to protect the places where we lived.

Wilderness was beautiful but misleading. The idea of going to the wilderness to get back in touch with nature was all wrong. The places where we needed to form close connections with nature were not in the wilderness but where we grew our food, heated our houses, took our daily pleasures. That was the place for me. The cities and suburbs, and the farms and forests that supported them, were the heart of the matter. Deal with that, and wild nature would more or less look after itself. Fail to deal with that, and wilderness would be left the long task of reestablishing itself over our ruins.

I did not develop an antipathy for wilderness that day. I still admire wild places, visit them on rare occasions, and believe we should preserve all we can. I did begin to see that preserving some land that is truly wild—that is, untouched by human activity to whatever degree of purity one wishes to specify—is of far less importance than learning to treat the greater part of the land well, to care for it. The two are not incompatible, but I began to suspect that truly protected wilderness will follow from a society that has at last worked out a healthy relation with its everyday landscape, with its productive forests and farmlands. And when that day arrives, paradoxically, wilderness will be discovered to have a much smaller spiritual and ecological importance than people once believed.

The more I have thought about this since then, the more I have been driven to the conclusion that environmentalism's fixation on wilderness is a mistake. Maintaining biodiversity, fostering healthy ecosystems—these things remain vitally important to me. They cannot conceivably be accomplished solely by wilderness or even mainly by wilderness—the amount of land that needs to be protected is simply too large. I am still adamant in my belief that a large part of the earth's surface must be left in *relatively* unmanipulated natural systems composed of a great complexity of species, not just the few we desire to harvest. Perhaps some completely wild places should be a part of that natural matrix, at every scale, for the sake of modesty and reverence, as Thoreau suggested. I am happy to grant that. We need *wildness* all around us, but the emotional appeal of total *wilderness,* of large regions untouched by humanity, truly is a construct of the industrial mind, an overreaction to an overcivilized way of life. If we can radically reform industrial capitalism and make it ecologically sound—an immensely difficult proposition—the concept of wilderness will lose its salience. Wilderness appeals mostly to city people. In the succeeding quarter century I have learned that when I work steadily to care for my home place and am engaged with natural complexity and beauty every day, the call of wilderness grows faint.

This revelation came to me on a mountain in 1975—or at least it began to come to me there. The same insight has occurred to others in the environmental movement in the past few decades, and several of them have explored it more fully than I have here.[2] It has provoked controversy, not to say consternation, because many seasoned environmentalists fear that questioning the value of wilderness means unlocking the door to the worst perversions of so-called wise use. This is certainly a legitimate concern, and it is why I oppose taking any land out of wilderness designation. But the most important thing I have learned since 1975 is this: having this insight and writing about it is easy work; doing something useful about caring for one's home landscape is hard work. But the rewards are great. That is what this story is about.

The next morning Chuck and I stomped back down to Aspen in high spirits, loaded our gear in his Plymouth Valiant, and drove east. Within a few months I had dropped out of Brandeis and begun working in a Boston suburb called Weston, on a community farm called Green Power. I was back on home ground, in the belly of the beast, in the suburbs.

Bill McElwain in his greenhouse, 1976. (Jerry Howard/Positive Images)

Chapter 1

Green Power and Land's Sake

My first day at Green Power Farm was a good one. I rode the Boston and Maine commuter train from Somerville to Kendall Green station in Weston. It was a crisp fall morning in October 1975. Arriving at Green Power through a back corner of the farm, I came upon Bill McElwain behind a large oak tree seeing a man about a dog. He was startled to actually see a man, me, emerging from a nearby subdivision. Suburban farmers learn to ignore their surroundings.

I recognized Bill because I had been introduced to him the previous week by my Brandeis advisor, Larry Fuchs, a resident of Weston and a fan of Bill McElwain. Green Power Farm was a community farm project run by the Weston Youth Commission, part of town government. I had arranged to work for Green Power one day a week and to write a paper about it for Larry for independent study credit. (I never wrote it—then again, maybe *this* is it.) Bill was then in his midfifties, a vigorous man with sharp, grizzled features, a hawk nose, piercing blue eyes, and sparse white hair. Bill was and is still a blithely positive anarchist whose operating philosophy in dealing with obstacles to farming the suburbs was, "It is easier to gain forgiveness than consent." The name Green Power itself was emblematic of the farm's era and purpose: this was a farm devoted to radical social and environmental change, and somehow it had sprung up in an affluent suburb. I couldn't wait to learn more about it.

Bill led me into the field and introduced me to three Picker Power volunteers, middle-aged women who helped harvest produce in the fall after the kids who worked at the farm all summer had gone back to school. "Brian," said Bill in a grand mock-courtly manner, "je vais vous presenter Madame Moon, Madame Mercuri, et Madame Fish!" It was like being introduced to a coven of matronly suburban nature worshipers out tilling the good earth in the dawn of the New Age. It turned out those really were their names. Did they realize they could have conjured a cone of astral power to descend upon Weston if they had simply joined hands in a ring?

Rather than enact a neopagan rite, we pulled carrots. The ground was cold and crusted from frost. We loosened the carrots with spading forks, pulled green and orange bunches and swished them in a tank of icy water to rinse off the dirt, and packed them in wire crates bound for free lunch programs and tenant union coops in Boston. I knew that a garden carrot tastes better than a store-bought one, but I was not prepared for the sweet crunch of a carrot in the New England fall. I couldn't believe it was really true that something so much better than what I could normally buy came right out of the ground. The morning was chill and my fingers were numb, but my skinny idealistic body was warm. Crows were calling over the adjacent woods of bronze oaks and dark green pines in the bright morning light, and within ten minutes I was hooked on farming. I ate and crated carrots all morning.

In a way the first few years of my learning to work the land were like that day. I have seen the same excitement seize other young people since—amazement that even in the late twentieth century you can find yourself farming. You've got your hoe and a few acres of dirt, and there you are. Naturally for many people this gets old after one season. But for some the experience grows richer and deeper each time around, not in spite of but by virtue of repetition—every year, memory and anticipation connect the moment to the eternal. When I ate that first carrot I realized that many of the things that had appealed to me intellectually about farming were attainable in practice. This was a relief, more than anything else. I foresaw what I would be doing for the next few years.

After lunch at Bill's kitchen table—the first of a thousand lunches I would eat there—he turned me over to his assistant, a lanky, energetic young woman named Susie Ashbrook. We drove Bill's trademark turquoise Toyota Land Cruiser to the sugarhouse, which stood next to Weston's Middle School. There we met a troop of Brownies and went apple collecting. The apples had fallen in an old orchard where a housing development had recently been built—the new residents' backyards were littered with heavenly fruit for which they apparently had no use. We ran around with the Brownies picking up apples, hauled a dozen or so bushels back to the sugarhouse, dunked them in a tank of water to clean them off slightly, and squeezed them in an antique press. Voilà: cider.

Like the carrots, the cider knocked me out. It was without question the best I had ever tasted. I think the secret was that most of the drops we collected were tart Baldwins. Once again, Green Power subversives had gone out into the unsuspecting suburbs, redeemed a forgotten resource, and magically converted it into something that surpassed what you could buy in the local supermarket. It was an idea of great power. Who could tell what other wonders this suburban town might contain, wonders that a little imagination and initiative could unlock? In a few months junior high kids would be boiling maple syrup from sap collected all along Weston's scenic roads. What other people fantasized about in an idle moment and dismissed as impractical, this guy McElwain was putting into practice.

After the kids left, Susie and I hosed down the cider press and drove back to Bill's with a trailer load of spent pomace. Bill's house, on the Boston Post Road in Weston Center, had a side yard choked with agricultural implements: assorted trailers, a manure spreader parked under a hemlock tree, an antique stake-body truck, a big pile of pumpkins in front of a few dozen boxes of winter squash covered with old carpets, a platform scale hanging from a tree stump with the prices of pumpkins and squash scrawled in green magic marker on the back of a cigar box and a coffee can full of change nearby, assorted stacks of scrap lumber, cardboard boxes, and wire crates. To many it was an eyesore, but the town had not made the permanent commitment to Green Power that would have included buildings to house equipment, so the farm ran out of Bill's yard and the neighbors just had to live with it. The door to Bill's house was always open, and people banged in and out 365 days a year, 12 hours a day, on the way to cultivating the far corners of Weston. You can count on this if you start a community farm: your house will become a community farmhouse, willy-nilly.

The last thing Bill and I did that autumn day, as the sun went down, was to hitch the manure spreader to the Toyota and drive it to a small orchard near the Green Power fields in the northwest corner of town. Bill wound the spreader among the old trees, spreading pomace. It was a perfect way to end the day, putting the spent apples back on the land as fertilizer, in an orchard no less. I was moved by this gesture of fealty to the golden ecological rule of return—it gave the day a nice feeling of completion. Pull carrots in the morning, press cider in the afternoon, return compost for the next harvest at the close of day. For me, an imagined way of life had suddenly acquired form.

That evening, I stumbled up the stairs to my third-floor Somerville apartment with a gallon jug of cider in one hand and a bunch of carrots in the other, beat to the socks but exhilarated. I wrote to my high school comrade Fred, "I really enjoyed the work—of course, conditions were ideal—cool, and fairly light work; and a beautiful morning in a big field—and me getting the college cleanness off my hands and mind. I'd have been happy digging fenceposts." For the first time in months, I fell asleep as soon as I lay down and slept soundly. Farming was the life for me. I began going out to Weston several times a week, neglecting my other courses. I was soon joined by my former roommate Brad Botkin, who had dropped out of school and was looking for something to do. He came to Weston with me one day and caught the agriculture bug, too. We slipped easily into the Green Power program, which seemed to have been expecting us. Within a few months Susie Ashbrook took a job directing a fledgling community farm in neighboring

Natick, so I took a deep breath, dropped out of Brandeis, and assumed her position. For the next three years I split this $6,000-per-year job either with Brad or with Sonia Schloemann, a recent Green Power graduate who found farming more attractive than the University of Massachusetts. We split the salary, but most months we both (or all three) worked full-time and then some because we liked what we were doing so much we would rather be doing it than anything else. That was my introduction to suburban community farming.

In those three years I learned basic agricultural skills: how to drive dump trucks and tractors, back a trailer, shovel shit (more than a skill—an art), burn brush, hang buckets and boil maple syrup, build sheds and makeshift greenhouses, nail apple boxes, prune apple trees, move irrigation pipe—and cultivate fifteen acres of vegetables. Mostly I learned the dawn-to-dusk tenacity that farming requires, which can be summed up as the ability to recognize what needs to be done and the willingness to keep doing it. What I remember best, though, is the sheer excitement of having gone from fantasizing about living on the land to actually farming. It was breathtaking. "I was riding down the field the other morning," I wrote to Fred, "working the levers on the spreader (spreading lime), looking around, when suddenly it hit me: 'Well. I'm really farming. I'm a farmer.'"

One of the great things about farming at Green Power was the variety of the work we did. Tasks that would have become mind-numbing if done day after day seldom had the chance to bore us before we were on to something else. This is, of course, the antithesis of the industrial ethic. Often we tackled half a dozen different jobs in the course of a day. Some were planned, some—like fixing the tool we were working with or fixing something else so we could begin fixing the tool we were supposed to be working with—were unanticipated. Some days we seemed to be working backward, getting farther and farther away from whatever it was we had set out to do in the morning. E. B. White once called farming an "interminable errand," and I can confirm that. Mostly it involves getting something unstuck. Even when we settled into some steady task for a few days, the demands of the shifting seasons guaranteed that we would soon need to catch up with something else. Good farmers relish the complexity of constantly shifting priorities among interdependent variables. It comes with the milk.

The farm year began with pruning an overgrown apple orchard we were trying to renovate. This charming work appealed to our sense of being in farming for the long haul, tied into the seasonal round. "I really enjoyed the pruning, though it was long, cold work over the snow, because I can picture them blooming in a few months," I wrote. Bill had hired an itinerant orchardist from New Hampshire to supervise the pruning. He paid him by brokering a massive pruning campaign in another old orchard that ran through the yards of the housing development on Applecrest Road, where we had picked up apples to make cider in the fall. Our part of the deal was cleaning up the mountain of apple brush that was generated, and it was here I learned that hauling brush—especially apple brush—is more work than cutting it. I learned to back a trailer, too, which cost one mailbox and part of a fence. It was also on this job that Brad and I spent an hour spinning our wheels on a sheet of ice at the dump with a fully loaded trailer behind us. After trying to activate the four-wheel drive by pulling every knob and lever we could find inside Bill's jeep, it finally dawned on us that locking hubs were located outside the vehicle. Thus we learned the routine skills that preadolescents grow up with on the farm. It did not come with the milk in our case.

We began syruping in February, although we had been working on the sugarhouse since fall. We had put in a new brick floor and hauled truckloads of refuse lumber from an old barn to use as kindling. The sugarhouse, built the year before, was a rustic board and batten building made of locally milled white pine, surrounded by mountainous woodpiles and makeshift sap tanks. It stood beside a new concrete fortress of a middle school, built in that seventies style of schools designed to withstand insurrection. This little wooden building, in which thirteen-year-olds were encouraged to pitch cords of wood into a roaring firebox and to tend boiling sap in a twin-pan evaporator, may well outlast the middle school. Weston's roads were lined with magnificent sugar maples, and as the weather warmed we took crews of students and decked the trees with taps and buckets. When the sap ran, we hauled it back to the sugarhouse in a collecting tank behind Bill's jeep, taking delight in stalling commuter traffic as we crawled over the hills of Weston with our sloshing load. In a six-week boiling season we reduced about ten thousand gallons of sap to a couple of hundred gallons of syrup that was eagerly purchased by Weston residents. The final product was not too fancy, but the town relished having buckets on the maple trees and middle-school students making syrup.

As the frost came out of the ground in March, the last spring snowstorms passed, and the days began to rapidly lengthen, our attention turned to our main enterprise: the farm on Merriam Street. For several weeks in April we hauled horse manure to the farm almost nonstop. Bill had gotten access to a pair of surplus dump trucks from the Highway Department. We hauled mostly from a large boarding stable in neighboring Wayland, which had a mountain of manure that went back years. Using one of the town's smaller front-end loaders, we kept the trucks shuttling all day, one person loading and the other driving. We dumped the manure in windrows at the farm, and once the ground was firm enough set about spreading it in strips across the fields, pulling an old horse-driven manure spreader behind the tractor. We also cleaned up piles at a number of private stables in the neighborhood of the farm, forking the manure straight into the spreader and then running it directly onto the field behind Bill's ubiquitous workhorse Toyota. I liked seeing how fast a couple of energetic people with the right tools (long-handled, six-tine manure forks) could fill the spreader by hand in a tight spot. I also enjoyed mastering the rhythm and angles of operating the machine loader on a big pile. There was a place for each.

In April, snow peas, onion sets, and spinach went in the ground. Early in May we planted the winter squash and pumpkin seeds in widely spaced hills, covering several acres. Bill began putting in a dozen or two rows of sweet corn every ten days with a tractor-drawn planter. Soon we were making the first plantings of small-seeded stuff, carrots, lettuce, beets, and seedbeds of collards, cabbages, and other brassicas, using the antique-looking Planet Jr. push-planter with its adjustable seedplates. By the end of May it was time to set out the tomato, eggplant, and pepper seedlings we had started in improvised greenhouses and to put in the first cucumbers, summer squash, and beans as the soil warmed. And it was past time to keep ahead of the weeds.

And so it went into early summer, with more plantings, transplantings, and cultivation. There was a nice rhythm to the successive plantings, many crops at different stages. It was all like an intricate dance. The greens burst from their beds so that by July fifteen acres were covered with crops. The amount of work was staggering. Bill was often out at four in the morning tilling ground for the day's planting, and Brad and I rolled in at six or seven. In the afternoon, Bill generally turned to administrative chores or to writing his newspaper column, while we worked on, sometimes into the evenings, trying to stay on top of it all. There were peas and tomatoes to be trellised, irrigation pipes to be set up, moved, and repaired when they blew out, and of course the constant pressure of cultivation, hoeing away the tiny pigweed and lambsquarters before they could gain a toehold. Luckily in May a few farm veterans returned from college to help out—an engaging coterie of McElwain protégées from previous years, several of whom would go on to careers in agriculture. Their motto was "Women Will Feed the World!" Good news for the world, but even better news for Brad and me. Thanks to Green Power, Weston was not quite the social wasteland it might otherwise have been for two guys in their early twenties.

By mid-July the weeds were slowing down and harvesting was picking up. Now we had big crews of schoolkids, some of whom were more helpful than others. Three mornings a week we boxed up collard, turnip and mustard greens, cucumbers, squash, and tomatoes and sent them into Boston for one dollar a crate. Sometimes the city groups we dealt with sent out a van, and sometimes we loaded up the jeep and a trailer and drove in to various churches, housing projects, and food coops in Roxbury, the South End, and Jamaica Plain. This was the era of antibusing riots and great racial tension in Boston, so we were acutely aware that we were involved in a bold and timely mission of good will. One particularly heady morning Brad and I dumped a load of manure at the site of a new community garden in Roxbury, just a few blocks from where a white man had been dragged from his car and beaten into a coma a week before. There we were in a truck full of steaming horseshit with "Town of Weston No. 6" engraved on the door. "If this goes awry, you know where they'll find our bodies," I said to Brad. But working for Green Power meant living a charmed life, and we and our manure were graciously received.

Green Power was tied in with a small but vigorous group of urban activists who were making connections between gardening, food, and community. It was a pleasure watching these energetic organizers in action at meetings, conferences, farmers markets, and feasts. To these people, Bill McElwain was a shining star. That this unlikely character had managed to get kids from Boston's most affluent and exclusive suburb growing fresh food for the inner city just blew people's minds. The food wasn't much by itself, but it seemed to suggest that anything was possible if we just kept working at it. Bill was a folksy, humorous speaker, and people loved hearing him exhort them to get children's fingers in the dirt and build more links between the city and the suburbs. Bill knew how to play the country bumpkin and say outlandish but oddly sane things in ways that grabbed people's attention. With Bill, you were never quite sure if you were listening to a New Hampshire dirt farmer or a courtly physiocrat with a little Tom Paine rabble-rouser thrown in for good measure. Picture Pete Seeger with a manure spreader instead of a banjo, and you get the idea. It was great fun for Brad, Sonia, and me to tag along to these gatherings and share in the glow of Bill's achievement.

By September, the kids were back in school and things quieted down on the farm. In a mild fall we were left with crops that were still yielding heavily and not enough hands to pick them—hence, Picker Power. On a Saturday toward the end of the month we cranked up the cider press, cooked up a big pot of beans, and called out our volunteers to harvest the pumpkins and winter squash. All morning people helped box squash and load pumpkins into the truck, and we rolled them down to Bill's driveway. There we set them on pallets, covered them with old rugs against the frost, sold them, and shipped them out. Except for the late carrot crop and a few collards that hung on after frost, that about did it for the farm year. We cut down the tomato strings, disked in the stubble, sowed winter rye, and turned our attention to hauling more manure, working on the sugarhouse, and making cider.

There were precious few apples to be picked up in Weston in 1976 because unfertilized apples are prone to biennial bearing and the past season had been a big one. Instead we bought culls from commercial orchards in Harvard and Bolton, towns farther west. Sometimes we took along crews of kids to pick up drops for ourselves, and sometimes we just had the orchard drop a few bins in the trailer. We pressed cider three afternoons a week at the sugarhouse, and Saturday morning on the town green or at Bill's house. Because the cider was made primarily from Macintosh and Delicious apples instead of Baldwins, it was sweet and insipid. Few of the orchards were growing the good old New England varieties anymore. But the cider was fun to make, and we didn't have any trouble selling it.

If the apples weren't strictly local, at least the boxes we put them in were. The winter before we had helped clear some white pines for a couple named Sy and Ellie Reichlin, who had an old barn and decided they wanted to restore the pastures suggested by the old stone walls on their property and keep sheep. Bill had the pines bucked into short lengths and hauled them up-country to a box mill in old Number 6. (The trip led to one of the most infamous incidents in Green Power history. Selectman Harold Hestnes received a call from a New Hampshire state trooper who wanted to know why a Town of Weston dump truck sporting 1955 plates was cruising down the turnpike with no lights. Just how this inquiry got routed to Harold was never quite clear to me, but it is a story both Bill and Harold still love to tell. It is enshrined in the mythology of our town—it celebrates Bill's audacity and marks the outer limits of official tolerance.) On rainy days we trimmed those boxboards on a table saw and nailed up apple crates. They were good boxes—I still have a few of them. My wife, Faith, uses one to store yarn that came from our sheep, which later grazed on the very pasture from which the pines were cut. So the circle is still unbroken.

My first year of farming was about over, and I was pleased with it; overjoyed, in fact. I was gratified that my life was now following the natural cycles of the earth and was free of the artificial orbit of academia. One morning in November I was back to hauling manure with Brad from Wayland, and it began to snow—it was shaping up to be one of the coldest winters of the century. I wrote to Fred, "This snow makes a nice wrap-up for my year's work here—seeing the farm closed down and left to lie—goodbye and keep cold—while we bring in manure for next year—anticipating the seasons while we move in obedience to them." Yes, it is a cliché, but it's a good one to live by.

During the dark months when the workdays ended early, we often stayed for tea in the McElwains' kitchen. There was a big oak table and a handsome six-burner wood range that soaked us in warmth after a day out in the cold. Bill, his wife, Katchen, or his daughters Louisa and Amy would serve Earl Gray or Lapsang Souchong tea, elegantly warming the teapot first with boiling water. We talked about the state of the world and about Green Power and what we would do next. In the 1970s, we took it for granted that industrial society was on the downslope. "This is the time when the new world has to miraculously surface in the midst of the disintegration of the old," I wrote to Fred. True, Weston didn't seem to be disintegrating very fast, but we assumed that energy shortages and other ecological shortcomings would shut down suburban growth before too long. Visionaries at the New Alchemy Institute on Cape Cod and a loose network of back-to-the-landers and organic farmers across rural New England were working toward a future based on solar energy and aquaculture, organic farming and composting, wood heat and simple living. We were part of that movement. Our mission was to develop a model of an alternative local economy right there in Weston. This would become a necessity if and when industrial society went bust and would demonstrate a healthier, more attractive way to live in the meantime.

Green Power seemed to be going from strength to strength. The farm had been steadily expanding, the maple syrup was the toast of the town, people lined up for the apple cider. Reliance on local resources had been reborn. So far it was but an infant, but there seemed to be ample room to grow in Weston. The town owned some two thousand acres of conservation land, and young people in town were complaining that there was nothing to do. We would give them something to do. We envisioned a year-round farm center that would work closely with the schools and might also offer a work-study program to students who wanted to take a year off before going on to college. We could produce food for school cafeterias and recycle food wastes in methane digesters. We could build a greenhouse and a barn and integrate livestock from chickens to fish, from pigs to grazing animals in a tight system. We could manage the town forests to produce firewood, lumber, and even food crops like hybrid chestnuts, black walnuts, shellbark hickories, and sugar maples. We could experiment with solar and wind power and maybe even tap some of Weston's small streams for hydropower. Winter evenings we discussed what other groups around the country were trying and spun out ideas of our own. "We're going to garden this town until it sags with fruit," I wrote to Fred at the end of my first year in Weston.

The best of it was, we were doing this in the suburbs, right in industrial capitalism's backyard. It wasn't necessary to flee to the edge of the wilderness to experience these alternative visions. By working in the suburbs, we would build a model where people could see it. Black Elk said that once you had a vision, you had to perform it on earth for people to see; and Wendell Berry warned that the vision was easy, the performance difficult.[1] Green Power was our performance. "I think getting our solutions into the mainstream is the job, not creating a counterculture," I wrote Fred. We weren't trying to work within the system in the usual sense of lobbying Congress: we were up to something much more subversive—involving people with radical ecological alternatives where they lived. We could strengthen our connections with urban gardeners and farmer's markets, maybe bring city kids out to our farm center, too. By involving children in all these productive activities we would make Weston a more vital community, instead of the mere residential enclave it had become. In our view, a bedroom community was a contradiction in terms, an impossibility.

That was our vision. But what with the farm, cider, and syrup, Green Power already had in place a year-round program that absorbed all the energy of its existing staff. Could we muster the political support to expand the program? This turned out to be a much more difficult performance than I anticipated. Green Power Farm was a product of the social activism of the 1960s and of Earth Day environmentalism. Founded in 1970, Green Power began as a voluntary effort to grow food for the inner city and turned into a town-run program for young people. But times change. By the late 1970s, Green Power's place in Weston was being called into question.

Green Power was the brainchild of one remarkable man, Bill McElwain. Bill was a Harvard graduate and had been a New Hampshire carrot farmer, a private school French teacher, and an urban low-cost housing activist. During a family bike ride in the fall of 1969 the proverbial light bulb went on in Bill's head. As he and his family rode down a dirt lane off Merriam Street, they discovered a large, weed-covered field owned by a Jesuit seminary called Weston College. Nearby was another field where a commercial grower had for some reason left a lot of cabbage to rot. Here were good food and good land going to waste in Weston, while people in the inner city were going hungry. Wasn't affordable food as central to people's well-being as affordable housing? A quick comparison of urban and suburban supermarkets revealed that produce cost more in the city and was of lower quality. The Green Power idea was born.

On Earth Day 1970, a group of Weston residents responded to an article Bill had written in the *Town Crier* and gathered on a pine knoll next to the idle field. Bill proposed to utilize suburban land and volunteer labor to grow food for the inner city. It was an audacious idea that fit the tenor of the times. Weston College made the land available, and that year a crew of young volunteers hauled manure and cultivated two acres of carrots, beets, cabbage, tomatoes, and summer squash, battling the witchgrass they had stirred up. When the crops began to ripen, they packed a van full of produce, and Katchen drove it into the Orchard Park housing project in Boston and simply gave it away. They arranged to deliver produce regularly to the project's senior center. The food was distributed free, although both city people and suburbanites made coffee-can donations to help with the costs.

From this spontaneous beginning Green Power Farm grew along two main lines: supplying fresh produce to the inner city and involving suburban young people with the land. Four acres were cultivated the next year, and the farm kept expanding through the seventies until it leveled off at about fifteen acres. An acre or two were devoted to a community garden of individual family plots that is still going strong today. At first Green Power's produce was given away to a number of city programs with whom Bill established connections. By the third year a standard price of a dollar per crate was established—still a fantastic bargain. Soon, food also began finding its way into the city by another route. Women from Orchard Park came out to Weston a few times a week to help with the harvest and in exchange picked vegetables for their own families. After a few years more people learned about Green Power and began driving out to pick for themselves, and a flat rate of five cents a pound was instituted. We later raised it to ten cents, still a very good deal. By the time I went to work at Green Power in the mid-seventies as much produce was leaving the farm in private cars as in big shipments of crates.

Besides providing low-cost food for the city, Green Power Farm offered Weston's young people a chance to do community work. About the time Green Power was founded, the town of Weston appointed a Youth Commission because of growing concern about teenage alienation and drug abuse. Showing good sense bordering on genius, the commission (or YCom, as we fondly called it) hired Bill McElwain as its project director. Suddenly, this old anarchist was a government employee. He organized bike trips, outings to the inner city, and a recycling center at the town dump. Shortly thereafter, the town acquired the land upon which Green Power was located as part of a townwide land-buying binge. Green Power was incorporated as a Youth Commission project, providing a salary for Bill, a budget for seed and fertilizer, and enough money to pay the young farmworkers small wages. This attracted more reliable help as the volunteer idealism of the sixties faded. On the other hand, becoming a municipal program meant that legally any income earned by Green Power had to be returned to the town general fund, and this imposed at least theoretical limits on Bill's freedom of action.

Green Power became, from Weston's point of view, primarily a program for children run by its Youth Commission. Every summer a few dozen teenagers and preteens worked several mornings a week at the farm, planting, hoeing, and picking. From Bill's point of view, of course, the educational aspects of the program were important but secondary—or maybe I should say, derivative. Kids should be given the opportunity to do productive things to improve the world, not simply have learning poured into them like so many empty vessels, was Bill's educational philosophy. They would feel valuable if their efforts were valued. Officially, all the children we worked with were at risk teenagers with incubating cases of middle-class angst. We did give a fresh start to a few kids who were at loose ends or even seriously messed up, but the truth is for the most part we got the best-adjusted, most highly motivated kids in town. For a few years we even received a small grant from the State Department of Mental Health—Division of Drug Rehabilitation. This helped pay my salary, so I was assigned to take care of the paperwork. I dutifully recorded all the children who worked at Green Power as "potential drug abusers," which I suppose all humans are. Although this was a bit of a stretch, the good liberals at DMH were happy to allocate a small part of their budget to a preventative program like Green Power. They believed in their hearts that if the kind of community engagement we were offering young people were more common in our society, most of the hard-core human misery the bulk of their budget went to combat with so little effect would not develop in the first place.

Certainly most members of the Weston Youth Commission believed this as well, but many other people in Weston did not. They thought the Youth Commission in general and Green Power in particular were a waste of money, not to mention a textbook case of creeping socialism that undercut all basic American values. Bill's cheerful anarchism did nothing to discourage this. His column "Thinking Small" in the *Town Crier* was enjoyed by most of its readers for its outlandish proposals—things like having middle-school students build affordable housing out of pine logs harvested from the town forest, so that teachers could live in Weston. You could never quite tell when Bill was serious and when he was just making a point. This didn't make life any easier for members of the YCom, who had to answer for what Green Power actually did. They dreaded coming to their monthly meeting and being informed that those log cabins were already half-built.

As we looked to expand Green Power into a model of ecological farming and local self-reliance, it became obvious that we were institutionally misplaced. The YCom supported Green Power but did not share our radical vision of an eco-community rising where a wealthy suburb had once stood. In any case, land management on the scale we were contemplating fell more properly within the jurisdiction of the Conservation Commission. The YCom was comfortable with Green Power as it was—the thought of its getting any bigger made them a bit uneasy. And they knew very well that the town would not support an increase in Green Power's budget. Bill McElwain's freewheeling style may have succeeded in creating a program like Green Power where no one else could have imagined it but wasn't going to be able to take it any further. Green Power had reached the limits of its natural growth.

Bill had a second heart attack in the spring of 1978 and went in for bypass surgery that summer. I was officially made acting director, but in practice Brad, Sonia, and I ran the farm together. We managed to keep things going, but it was a hard growing season. Far from taking on new challenges, Green Power was struggling to stay afloat. "It's been a long, hot, troubled summer, and vegetables are boring," I wrote to Fred in August. Brad and Sonia weren't much interested in taking directions from me, and when Bill returned in the fall I discovered I wasn't much interested in taking directions from him. I also discovered that Bill no longer had much interest in expanding Green Power—he, too, was comfortable with it as it was. Bill's boundless creative energy was beginning to turn toward his land in New Hampshire, where he envisioned a larger, rural-urban cooperative project that eventually became Nesenkeag Farm. Green Power had become rootbound: if it was going to grow into something bigger, it needed to be transplanted to a bigger pot.

Green Power was self-limiting. Weston would tolerate Green Power as a child, but never accept it as an adult. Eventually, when Bill left the scene, the town was going to kill the whole thing. This became ever more clear as the property tax revolt of the late 1970s set in. All the wonderful new projects we had dreamed of, which seemed no more miraculous than what Bill had pulled off already but merely extensions of the same philosophy—the orchards, greenhouses, fishtanks, and tree crops—were never going to be funded. To do more of those things, we had to find a way of doing business that went a lot further toward covering its own costs. This seemed especially true if we expected other towns to follow our lead and try similar things, which is what would make our model meaningful.

Green Power was a bold pioneer but not the most sustainable model for a community farm. It wasn't even an adequate response to the problem of making good food available to the inner-city. It was a decent act of charity with which the burghers of Weston were perfectly comfortable. Having mechanisms to get good food to people who are indigent is a worthy goal, but lowering the price of food is no way to improve the lives of working people in general. It is true that fresh produce remains more expensive and harder to find in poor areas like Roxbury than in affluent ones like Weston. But the main reason poor people can't afford good food is that they are poor, not that food is expensive. We don't need to lower the cost of food any further in this world, we need to change our economic and political systems so that not so many people are trapped in blighted communities where they have so few opportunities to eat and live well. I don't know any easy ways to do that, but I do know that the big problem with American agriculture is not that food is too expensive, but that it is too *cheap*—and very little of what *is* spent on food returns to the farmer. Most of it goes to the companies that manufacture and pour on the syrup and grease along the way.

Really good, healthful food is culturally undervalued by almost all segments of our society, rich and poor alike. What we need are consumers who make decent livings and have the sense to buy good food as directly as possible from the farmer, and farmers who make decent livings by growing and selling good food, not just raw material for the food industry. The low-cost production of food as a mere sweetened fuel drives us toward further concentration into ever-larger farm enterprises that employ fewer people. We would benefit more from an increase in nearby farms that involve more people. If it couldn't generate more suburban food production than what the wealthy were willing to give away to the poor, the Green Power approach wasn't going to get us very far. In my view, sending a few boxes of cut-rate collard greens from Weston into Roxbury every year was a good thing, but it wasn't really addressing the problem of how to build a workable local food system.

We needed a more flexible model. An approach that freed community farming from the financial constraints of direct municipal control and earned more money than Green Power would go further. It ought to be possible to keep the ideals of working with kids and treating the land in ecologically sound ways and still sell products at a price that paid for a larger part of the operation. A small subsidy from the town on the order of Green Power's budget could then be leveraged into a much larger program. This would create a model that could be replicated elsewhere, especially in towns less wealthy than Weston.

I was learning that running a community farm required good politics as much as good agronomy. Here was Weston, a suburb with two thousand acres of commonly owned open space. How did the town come to acquire that land, and for what purpose? Could a way be found to involve young people with conservation land that resonated with Weston's sense of itself and yet moved in the direction of meeting social and environmental goals that had not yet occurred to most residents? What kind of place was Weston, anyway?

Weston

During my first years at Green Power I lived along the old Boston Post Road, in Weston Center. Weston has a pleasant center. At the heart of the town, a long celery bowl–shaped green is encircled by the Town Hall, the old library, the Josiah Smith Tavern (known locally as the Jones House), the central fire station, and two handsome stone churches. West of the green along the Post Road lies Weston's small business district: a supermarket, the post office, several banks, a dozen real estate and insurance agencies, and a handful of offices and shops. The center is sandwiched between a rocky ledge to the north and a red maple swamp to the south, leaving room for only two side streets. Beyond the business district to the west lie the town cemeteries, another antique tavern and another church, and the village houses along the Post Road. Many of these are typical New England two-story, white-clapboard frame structures, complete with third-floor attics and dormers, black or green shutters that never shut, and Greek-pillared front porches. East of the town green on the Post Road and Church Street stand more sturdy old houses, more churches, more trees. By the town green grow shapely lindens and horse chestnuts. Until recently, the Post Road was lined with grand sugar maples along much of its length.

An altogether charming place, Weston Center, except for the cars. One bright May morning twenty years ago I counted twenty-five cars passing along Church Street every minute, but only three walkers in two hours—a ratio of a thousand to one. There are more pedestrians and cyclists today, but more cars, too. Weston continues to suburbanize and endures an increasing cross-flow of traffic from neighboring towns—even as they endure us. But in spite of the frenetic automobiles and encroaching development, Weston retains much of the feeling of a New England town. And it behaves entirely like a New England town.

In order to make sense of this story, readers need to understand what a New England town is. Some will know this in their bones, of course. A New England town is not an urban place (as in the usual meaning of *town*)—indeed, most New England towns are predominantly rural, although many are becoming suburban. A town in this region is about the same as what is called a township elsewhere, except that townships I have lived in provide only a bare-bones municipal scaffolding that lacks most of the sinews that tie New Englanders to their towns. New England towns are real places—Westonites think of themselves as living in Weston, and when they cross the town line, they know it. Towns have a solid physical structure and well-exercised political and cultural musculature to go with it. Fundamentally, they are small, bounded, self-governing communities. They are typically about five miles across—the size can be larger than that in lightly settled areas, but as they approach ten miles across, the town notion begins to feel a bit stretched. Through most of the New England countryside there is no unincorporated territory between towns—you are in either one town or the next, even when standing deep in the woods. The selectmen of neighboring towns meet every few years and perambulate the bounds, as they have for centuries.

Towns usually contain a few thousand people—those that have been industrialized or suburbanized may have more. Weston has more than three thousand households and ten thousand inhabitants. Once towns are this populous the town concept of direct democracy among neighbors who know one another can get cumbersome in execution. At this size, town meeting government works only because typically fewer than half of us participate in town affairs. Only a few hundred are involved to any great extent, serving from time to time on the various elected and appointed boards and committees that govern the town. The others might as well be living in the suburbs of Phoenix, and we don't pay them any mind.

Near the center of most New England towns sits the village, or center. The center is not a separate political entity with a fixed boundary of its own but simply the nucleus of the town. New England villages with their town greens, maple trees, modest civic buildings, and spare churches are full of native charm. They often look as if they sprang directly from the early nineteenth century, the halcyon days of the Republic—and many town centers were indeed elaborated from very rudimentary beginnings during that era, as commercial activity quickened. In many cases, however, this antique feeling is a more recent artifact, a deliberate concoction. Weston Center owes a good deal of its charm to the dedicated efforts of a later generation of citizens who moved to the town during the late nineteenth century. By the early twentieth century Weston was dominated by a class of wealthy Bostonians, several with fortunes made in textiles. Some of these merchants and manufacturers built houses near the center and commuted to their Boston offices on the Central Massachusetts Railroad; others bought outlying farms and built country estates to which they retired during the summer months, spending most of the winter season in Cambridge or on Beacon Hill in Boston. Weston a century ago was already a country getaway.

These turn-of-the-century newcomers were determined to keep Weston an attractive New England town or to make sure it became one where it wasn't already. The center was a string of stores and stables jammed along both sides of a narrow upland, hemmed in by swamps—too ramshackle by far. Ideas for the betterment of the center were bruited about for some years, until in 1911 a town improvement committee was appointed at town meeting. Under the direction of the noted landscape architect Arthur Shurtleff, the town purchased a swamp behind the old town hall at the east end of the center, consigned the intermittent stream that traversed the swamp to a pipe beneath it, built a new road around it, and planted grass and trees over it. Thus they created from whole cloth a Weston town green, or common, a New England village fixture that Weston had previously lacked. Next they knocked down the old town hall that stood hard against the Post Road with the swamp at its back and in 1917 substituted a handsome red brick town hall with august white pillars on the far side of their new town green, lending spaciousness and grace to Weston Center.[2]

That generation of improvers did its job well. Besides creating a town common in the center, they saw to the planting of sugar maples along all the main roads because these too were required of a proper New England town. Never mind that sugar maples were hardly known to grow in that part of New England before. It is instructive to learn that much of what we view today as authentic native charm was itself fabricated during an era of nostalgia for an idealized New England of a still earlier period. Whatever its social merits, that movement to spruce up the rather shabby rural character of Weston proved an esthetic success and left later generations something worthy to build upon. It marked the beginning of Weston's movement to reclaim the commons.

The next generation faced a greater menace in protecting the character of Weston: themselves. Weston lies west of Boston just beyond Route 128, the city's inner beltway encircling such older suburbs as Lexington and Newton. After World War II, the attractive rural town straddling the recently completed Massachusetts Turnpike became an easy commute. Many of the large estates that had dominated Weston since the Gilded Age had fallen on hard times during the Depression, and large tracts began to come onto the real estate market. New housing developments went in throughout the town. Soon, both the newcomers and a few remaining scions of the old guard began to fear that the open, rural character of the town was in danger of disappearing before their eyes.

Commuters were their own worst enemy because every purchase of rural ambiance undermined the rural economy sustaining it. This contradiction became the central dilemma of suburbanization. The suburban drive turned rural character into a nonrenewable resource that could only be consumed, rising steeply in value as it became more scarce. Half a century later this dilemma has not been solved and has spread far and wide to take in more and more of the countryside, in a chain reaction of development and flight. Large lot size, that sine qua non of suburban zoning bylaws in the postwar era, simply makes the situation worse by consuming land all the faster.

Suburban sprawl is a true "tragedy of the commons,"[3] by which the unbridled selfish interest of individuals (in this case the desire to own their own piece of the country) consumes and destroys an unprotected common resource (in this case the beauty and integrity of the rural countryside). Fortunately, a few places such as Weston have found at least a partial response to this scourge, which is to purchase community-owned conservation land—that is, to defend the commons.

The idea of towns acquiring municipal forestland is actually more than a century old in New England, going back in Massachusetts to enabling legislation passed in 1882. That was the era of the birth of the conservation movement and of grave concern about the future of the nation's timber supply. This is no surprise given the ragged shape of the forest from Maine to Minnesota in the wake of the nineteenth-century logging boom. In response, conservationists and foresters strongly encouraged the establishing of town forests and the replanting of trees. The town forest movement peaked in the 1920s, then gradually declined. It became obvious that New England was reforesting on its own, and the long-term rewards of owning and caring for either young pine plantations or scrubby natural regrowth were too uncertain to interest many towns. For a time, community forestry seemed superfluous to nature's more sweeping reforms. After World War II, however, the advent of suburban sprawl and the rise of the environmental movement brought renewed interest in protecting open space. In the late 1950s and early 1960s, legislation was passed in Massachusetts that enabled towns to form conservation commissions to inventory open space and to set up acquisition funds to purchase land. The town conservation land movement was born, and Weston was among the leaders.[4]

The greatest success in saving land was achieved in places where coalitions formed between ardent environmentalists among the newcomers and political insiders among the old-timers who wanted to preserve something of the town they had known. People of contrasting political ideologies learned to work together effectively on this issue. Of the many activists who worked for conservation in Weston, one stands above the rest: William Elliston. A British-born physician, Bill Elliston and his wife, Harriet, moved to Weston in 1937. After the war, Dr. Elliston became involved in town affairs and led the conservation effort for the next three decades, serving on various open space committees, the Planning Board, and the Conservation Commission. He was also instrumental in organizing the Weston Forest and Trail Association, a quietly effective nonprofit group working with the town to protect open space and to develop and maintain a townwide trail system. Land preservation set off down a long, winding trail. Ad hoc committees sometimes evolved into permanent town boards and sometimes came to nothing; careful planning seemed to go nowhere only to be followed by abrupt action behind the scenes when the chips were down—all the usual political detours. But Elliston and a few others saw it through and achieved a notable result.

Weston acquired its first few hundred acres of backland in the late 1950s and early 1960s. Then preservation efforts moved slowly for a time, as real estate values steadily rose. A powerful faction of the political establishment in Weston was always ready to argue that land acquisition had gone far enough—it was costly, imposed a management burden on the town, and removed property from the tax base. Some held that the town simply had no legitimate business owning so much land. But as the environmental movement blossomed and development ground on, support for conservation land grew. Conservationists argued that the purchasing of open space would ultimately cost the town no more in taxes than the increased services that would be required if the same land were developed and began producing schoolchildren. In 1972, town meeting authorized a bond issue of $2.8 million for open space. The Conservation Commission got busy buying land in every corner of Weston.[5]

I arrived in Weston just in time to witness the tail end of the great wave of conservation land buying, when the Weston College land came on the market in 1976. This 145-acre parcel lay between Merriam Street and Concord Road and belonged to a former Jesuit seminary. It was a rolling landscape of small fields, orchards, and diverse woodlots, with a handsome skating pond in the middle of it all. It was, and is, a charming piece of land. Green Power Farm occupied adjoining fields that had already been purchased by the town. My introduction to the politics of land protection and the workings of town meeting had begun.

Weston College was asking $1.4 million and gave Weston first crack at buying the land. The town had already spent its conservation fund and would need a fresh appropriation. A partial development proposal won the support of the Planning Board and selectmen: it called for condominiums to be built on one-quarter of the property, while the other three-quarters would be preserved. The town had to come up with half of the purchase price. This was a prudent proposal, and I would surely support it today. At the time it was vociferously opposed by many in the neighborhood, including us Green Power farmers, who could not bear to see condos rise in the midst of "our" beloved apple orchard.

At a stormy town meeting in January 1977, the selectmen's motion that the town appropriate $700,000 to buy half the college land was defeated by an odd coalition of those who wanted the town to buy all the land and those who didn't want us to buy any of it. A motion to buy the entire parcel was made but foundered on the reef of a series of conflicting amendments offering different bids to Weston College. The conservation vote had splintered and every option was defeated, leaving the land up for grabs. It was a classic example of a town meeting disaster in which the whole thing crashes and burns. The way seemed clear for a conventional development of single-family houses. This is exactly what happened a few years later after the defeat of a similar proposal for a piece of land on the other side of town: we blew the condo convoy out of the water only to wake up and find ourselves gazing upon a fleet of houses the size of ocean liners, anchored over the whole lot.

The tale of the college land had a happier ending. A matching grant was secured from the state to help with the purchase of the entire 145 acres, and Weston approved its half of the funds at town meeting in May. This time it was a wild night outside the hall. A freak nor'easter loaded newly leafed maples and oaks with six inches of wet snow, and by evening branches were crashing down everywhere, blocking roads and wiping out power lines. The high school auditorium went on emergency power as the purchase of the college land was moved to the floor. Dissent was raised by some of the real estate interests in town, who argued that too much conservation land had already been purchased in the northwest quadrant of Weston, and that this was unfair to those living in the other quadrants. This inspired one of the finest town meeting speeches I have ever heard.

After the realtors made their pitch, a silver-haired man was recognized by the moderator and approached the podium. "Hanson Reynolds, Colchester Road," he began, "member of your School Committee." A standard opening for a town meeting speech. Its purpose is to remind everyone that the speaker is a resident, is solid and respected enough to have been elected to a town board, and isn't just an eccentric whose town meeting diatribes are eagerly anticipated but not taken seriously. "I see I am also a resident of the northwest quadrant," Hanson observed. The audience shifted forward in their seats, sensing that they were being set up. This quadrant argument was so contrived it seemed to be begging for it. "I had never thought of myself as living in a quadrant before," Hanson continued. "I had always thought of myself as living in Weston. But I see from the map that I do live in the northwest quadrant, and my family and I often walk in the woods on our wonderful conservation land in Jericho Forest—in the northwest quadrant. But we also go to the dump—in the northeast quadrant. Sometimes we go sledding at Pine Brook golf course—in the southeast quadrant. And the kids go to our fine schools in the southwest quadrant. All this takes place within small boundaries."

Hanson paused and then declared in ringing tones, "And so I say to you: Ask not what Weston can do for your quadrant. Ask what your quadrant can do for Weston!"

That brought down the house. When the roar of laughter and cheers subsided, Hanson concluded with a few lines from Thoreau about the value of wildness and sat down. The best town meeting speeches are brief. The motion to purchase the college land passed handily, and the quadrant issue has not been raised in Weston since. That is the way affairs are governed in New England towns.[6]

With the meeting over and the college land safe, we headed home, which wasn't easy. I was riding with Bill McElwain and his family. Wellesley Street leading north to the center was cut off by fallen trees. We went the long way around on Highland Street, dodging downed branches and power lines all the way. I remember standing on the Post Road in the utter dark of the storm and the blackout, listening to the steady pop and crash of one branch after another back in the woods, feeling pure elation to be alive in such a world. The oak tops that fell in that storm were recognizable for years afterward. Their bark turned solid black, and they wore an outraged expression that snow could have come so late, catching them with their leaves already out. It is risky to stand at the cutting edge of evolution.

By the late 1970s, some 1,700 acres of town forest and conservation land had been purchased in Weston, and a bit more has been added since then. The same movement was under way in neighboring Concord, Wayland, Sudbury, and other suburbs. Towns approved open space plans, passed substantial bond issues to buy land as it became available, and went to work saving what they could. It was not possible to control the shape that rampant development took by this piecemeal approach, but it was at least possible to soften its impact. Counting land set aside for various municipal purposes and land owned by the Forest and Trail Association, today some 2,250 acres of Weston are community-owned—better than a fifth of the town.

Towns have followed differing strategies about what lands to protect. Lincoln, lying north of Weston, is renowned for the handsome stretches of farmland that have been preserved along its roads, lending that town an open, agricultural character. Weston followed Dr. Elliston down a less-trodden path into the woods. Deals were struck with several of Weston's largest landowners by which forested backlands were sold to the town at reduced rates, while the owners retained the street frontage to sell on the rising land market as they pleased. In fact, the value of those frontage lots was instantly enhanced by the conservation land at their back. As a result, Weston has preserved as much acreage as Lincoln, but most of it is invisible from the road. Instead of pastoral viewscape, Weston got a hidden wilderness. Finding it is a local secret: you have to stroll down a cul-de-sac past grand houses, spot the nearly invisible trail marker, and go boldly up a narrow right-of-way along somebody's driveway—disappearing into the woods between their swimming pool and the neighbor's tennis court. One such side-street on the edge of Jericho Forest is named Elliston Drive.

There was wisdom in this strategy because it protected the most acreage for the money, much of it in large tracts. It did leave our generation with the problem of preserving at least a few of the remaining highly visible, attractive, open parcels along the roads. These are now worth an inordinate amount of money—thanks in part to the success of conservation around them. With diligence, once all this land is either built or saved we will probably end up with close to one-quarter of Weston in commons. And that presents another challenge: how best to care for all this open land.

Land's Sake

Weston's picturesque center, a few winding roads lined with stone walls, a handful of hobby farms, and a couple thousand scattered acres of overgrown conservation land are all that is left of a former agrarian town of farm lanes, hayfields, pastures, and woodlots. The rural landscape has now all but vanished beneath suburbia on one hand and resurgent forest on the other. Once I had lived in Weston for a year or two, I began to stumble on the remnants of this world that was. Green Power Farm lies about a mile north of the center, a pleasant walk up Concord Road. After a while, I discovered another way to walk to the farm, by going through the woods. I could circle west into Jericho Forest and then follow an old lane that would eventually bring me out at the back of our cornfield. I enjoyed passing through the woods to reach the farm, emerging from the cool morning shelter of the trees into the bright sunlight and broken earth to take up the tools of husbandry. This daily journey seemed to place agriculture in its proper context as something tentative on the earth's surface, contained within the eternal forest.

Jericho Forest is a large collection of town forest and conservation land in the northwest part of Weston that was acquired piecemeal by the town over several decades. Jericho proper comprises about 545 contiguous acres, which with a few adjoining tracts makes almost 900 acres of conservation land broken only by lightly traveled roads. It took a long time to get to know this area. At first it seemed to stretch toward the west and north without end. This was partly because the trail system in Jericho was so loopy that I could walk all day without passing the same way twice. Conversely, on some early walks no matter which way I turned I kept returning to the same intersection many times, as if I was in an enchanted forest with no way out. Like New England roads but on a smaller scale, the trails in Jericho must thread their way among glacial till knobs, gravel lobes, peat bogs, and kettle holes. They skirt swamps along the higher ground and cross them on ancient causeways. There are no high hills in Jericho, no outlooks, no discernible lay of the land at all. Much of it is a large swamp sprawled among an irregular collection of low kames and eskers that go wandering in no particular direction. Entering Jericho, I seemed to pass through the mists into a primeval world.

Jericho Forest was an unexpected discovery, an "unforeseen wilderness" in Wendell Berry's phrase. It didn't seem as if such a wild place ought to be there in the middle of the suburbs. Just over the line in neighboring Wayland was an equally endless housing development, but it was easy to keep that knowledge on the far side of a certain protective mental boundary as I walked the quiet trails in Weston. Hoeing in the cornfield, I could hear white-throated sparrows, oven birds, and veeries singing back in Jericho, and sometimes walking home in the evening through the woods I would stop, spellbound, to listen to a hermit thrush singing vespers. Deep woods birds, northern birds. My good opinion of Jericho was confirmed one May morning when I was attacked by a furious goshawk guarding its nest. Breeding goshawks in Weston? I exulted as I knelt in the pine duff, checking to see if I still had my hair—this was wilderness incarnate.

There is some truth to this impression of Jericho Forest as a hidden wilderness. Jericho does show that if blocks of forest can be kept large enough, even in the suburbs, they will support interior species that do not thrive in smaller, fragmented woodlots. Something else was going on within this wild woodland, however, or had been going on there once. To begin with, there were dozens of stone walls to be seen. They ran everywhere, over hills, around the edges of swamps, often along both sides of the trail. The Weston Forest and Trail Association published a fine set of trail maps that carefully recorded many of these walls. The sheer density of walls in Jericho and the intricate pattern they made were deeply thought provoking. A few very long walls running straight over hills and through swamps suggested an ancient grid that the other walls subdivided. Most of the walls were short and crooked, stopping and starting again, parceling the land into interlocking pieces whose working relationship was now obscure. The force that had obscured and all but obliterated that pattern couldn't have been more plain, however: it was the forest. This land had all been cleared and farmed, and then the forest had taken it back. Jericho Forest was not an undisturbed wilderness at all, it was a *much*-disturbed landscape of agricultural abandonment.

Now that I appreciated the effort involved in farming a little myself, I began to grasp the scale of what had happened in Jericho and a thousand northeastern forests like it. Here generations had worked without machinery, with only oxcarts and axes, crowbars and spades. The labor sunk into those ubiquitous walls and the former fields they enclosed was astounding. Farmers hadn't merely pushed back the fringe of the forest; nearly everything in Weston had once been cleared and farmed. In time I learned that by the middle of the nineteenth century only about 10 percent of the land in towns like Weston remained covered in woodlands; the rest was scratched with plows and mowed with scythes or grazed to the roots. The forest had been cast off all but a few rocky hills and steep ravines, and fields and fences held sway, apparently for good. Then, after being painstakingly cleared, this farmland was abruptly reclothed in trees. And not so long ago, either. Old-timers in Weston told me that many of the fields in Jericho Forest had been given up during the 1930s—they could still remember when the area was open and farmed. That people could live to see deep forest where they had cut hay and pastured cattle as children was amazing to me. I was walking among the ruins not of ancient history, but of living memory.

Ditches marked Jericho Forest as thoroughly as walls. Down the middle of every red maple swamp ran a straight channel, often with several cross ditches feeding it. These ditches connected from one swamp to the next, forming another intricate network that had once drained the entire watershed from top to bottom, going on for miles. I later learned that many of these wetlands hid tile drains as well, buried two feet deep in the muck. The ditches had all silted in; the swamps lay flooded much of the year and grown up in red maple, elm, highbush blueberry, winterberry, northern wild raisin, marsh marigold, and skunk cabbage. But it was clear that at one time dozens of farmers had gone to great trouble to ditch and drain their lowlands, just as they had hoisted thousands of stones to clean and fence their uplands. The fall was so slight that every farmer must have been obliged to clean his ditches or else hundreds of acres of hay meadows above would have backed up and reverted to swamp. I imagined the immense collective social pressure on those owning land below—and later learned there was a law allowing upstream farmers to cut drains over their neighbors' property if need be. Against the uplands were watering places for stock, and along the main brooks were sites of small sawmills and ice ponds. In brief, the watercourses in Jericho were no more purely natural than the vegetation. The present streams and forests had wild roots and a wild crown, but for better or worse the body of the land had been reshaped by human hands during several centuries of husbandry.

The old lanes upon which I was walking had once been part of a third network by which the owners and their teams worked the alternating uplands and meadows. If I looked hard enough, I could find the gates and the faint tracks that once led into these fields, where cattle and haycarts had passed. Some of the lanes had formerly been maintained as town ways—once winding through open meadows and fields, now lying deep in shade and shrouded in fallen leaves. Here generations of farmers had worked off their highway taxes hauling gravel, cleaning culverts, building causeways, and mowing weeds. A powerful cultural ideal had reordered this landscape, and then the forest had grown back over it, like some lost Mayan civilization. The difference was that in Weston much of the collapse had occurred within memory—although it seemed just as thoroughly forgotten by most of the present inhabitants as if it had happened a thousand years ago.

The land in Weston, farmland and forest alike, was a product of human as well as natural history. The line between our cornfield and the adjacent forest was not as sharp as I had supposed. I gradually became aware of not only how broad Weston's conservation land was, but how deep it was. New questions began taking shape. How exactly had this land been used, and why had it all been abandoned after such an arduous investment? How had the land itself been changed by its sojourn under cultivation? Had the soil been depleted, and did this contribute to the decline of farming? Was the forest that grew back degraded in some way, or merely different, or essentially the same as what had gone before? What exactly had gone before, anyway? How had Native Americans treated the land here for thousands of years before our culture imposed itself upon the scene? It seeped into my mind that if we activists hoped to make ecologically sustainable use of this land, we needed to know more about what it was we meant to sustain. The land's promise and its limitations were all grounded in its past. Sustainable use wasn't just something we could concoct out of wild air, earth, fire, and water, like alchemists with a little eco-lixir. We had to know what we were working with.

There was a powerful message in those mute walls and ditches. This seems obvious to me now but hearing it at first required a subtle shift in how I held my head. I had been thinking that what we were doing at Green Power and all we proposed to do as the environmental crisis deepened was something new. The journey from the overindustrialized present that the suburbs epitomized toward a decentralized future would require radical appropriate technology and ecological thinking. Visions of solar greenhouses, aquaculture, wind turbines, human waste recycling, intensive organic gardening, and forest farming with edible tree crops constantly danced in my head. Our history, I had assumed, was little better than a long catalog of environmental abuses. I took it from Henry Thoreau, Aldo Leopold, Paul Sears, and many others that agriculture had been an ecological scourge for a long time, maybe since the very domestication of crops. Certainly since the advent of the plow and the taming of the sheep. Not much to be learned there except what to avoid. I was not one to romanticize the agrarian past. We needed to go back to square one, that is, to wild nature, and design anew.

Now it dawned on me that a form of decentralized, small-scale agriculture had once been practiced in Weston. Like it or not, this legacy prepared the ground on which we had to plant, both underfoot and in our heads. The past was inescapable. The butternut squash we grew, the maple syrup and apple cider we made, the condition of our forest and the character of our soils—both the land itself and most of our ways of farming it came directly from that tradition. We could not have cultivated our fields, organically or otherwise, had the stones not already been removed—and by cultivating it, we committed ourselves to removing more stones as the land continued to heave them up into our hands. At Green Power I had already been exposed to some of the ecological virtues that local farming once possessed, tried and true ways of fitting a wide range of crops into the topsy-turvy landscape and climate. No doubt rural New England had also suffered from serious ecological flaws; I strongly suspected that much of the reforested land in Jericho should never have been farmed and had probably been deeply if not permanently scarred, and that the forest had grown back greatly diminished. But in spite of these misgivings, I came to understand that we should be working not on the creation of a completely new way of living in this part of the world, but rather on the recovery and reform of a flawed but still worthy, older way of living here. We were not up to anything new. We were heirs to a wounded agrarian tradition.

Following a long decline, in just the past few decades these ways of living from the surrounding land had finally been abandoned—their virtues and flaws alike had ceased to hold any meaning to modern residents. After passing centuries as slowly evolving farming communities, New England towns like Weston were being turned more or less overnight into rootless bedroom communities in which children grew up bored and angry in spite of boundless affluence, and their parents wondered why. Could it be at least partly because neither generation any longer had meaningful ties to the land they inhabited or to any cultural tradition in that place? Is it possible to have healthy families without healthy communities, and is it possible to have healthy communities without an ongoing relationship to the land? It seemed to me that even in an urban and suburban society, those links had to be retained and rebuilt. The suburban quest for rural character was a clumsy expression of that desire, but like so much in our society it arrived only at superficial fulfillment; it was an empty shell. Children rebelled quite rightly against the shallowness of suburbia, but after their period of youthful seeking what would they return to? Not to the homeplace, but to the same placelessness being replicated across the continent. Only vestiges of the rural past remained in Weston, not enough to welcome anyone home. The history of the land was generally regarded by parents and children alike as quaint at best, if not entirely forgotten and unknown.

The essential task before us was not merely to foster ecologically sound land use methods. That was important work, but it could be accomplished only given a change in prevailing attitudes and economic climate. The essential task was to bring about a cultural transformation, and that was going to be a long, hard pull. People needed to develop a deeper sense of place, a deeper engagement with place if we were going to prevent suburbanization from rolling on for decades until the oil finally ran out, doing immeasurable damage along the way. We needed to convince people that understanding the places where they lived, caring for the places where they lived, and having a practical connection with the places where they lived were worthwhile goals in life. The community's land had to be more than something to absorb effluent, and the community's history had to be more than something to absorb antiquarians.

The daunting scope of what we were really up to began to sink in. Our job wasn't so much to invent a new agroecology as it was to recover, revive, and in the process radically reform a traditional agrarianism that was nearly gone. With almost a quarter of Weston now in community-owned conservation land, we had an opportunity to work at this on a significant scale. But would Weston want to use its conservation land in this way?

Taking care of Weston's open space was the responsibility of the Conservation Commission, a board appointed by the selectmen. By the late 1970s, the commission was beginning to ponder land management. After the purchase of the college land it appeared that the era of acquisition in Weston was largely finished, and attention turned to the question of what to do with all this land. What, if any, management did it require? Weston had no land management arm. In 1977, the commission designated a land use study committee. I served on this group, along with several others, including Julie Hyde, a young woman who had grown up on one of the last farms in Weston and was about to embark on a career of growing hydroponic tomatoes behind her house (although she may not have known it at the time). Also on the committee was Doug Henderson, a diplomat who having retired from the Foreign Service had returned to his boyhood home in Weston and become involved in town affairs. It was a group that cared deeply about Weston's land and had a working knowledge of it.

We met several times and made our recommendations. Weston's conservation land, we said, was an asset to the town that should not be allowed to depreciate through neglect—that seemed like the kind of language that would appeal even to people who weren't environmentalists. We recommended first an inventory of the land, which would characterize the various parcels ecologically and assess their needs and potential. Then management plans would be generated, featuring uses of the land that involved young people. We had in mind more farming projects on the open parcels, reviving orchards, planting sugar maples, experimenting with blight-resistant American chestnuts and other nut trees, enhancing wildlife, and trying some very limited forestry. Finally, we recommended that the Conservation Commission hire a staff person to carry out this program. It was an ambitious proposal.

The commission approved our recommendations in principle; in practice, nothing came of it. This was an era of property tax revolt, town budgets were being trimmed, and Weston was in no mood to bring on a new staff person. The time was not right for a town-run land management program. There was some thought that Green Power might take on these projects, but the Youth Commission wasn't eager to expand Green Power, either. Still, the effort of the study committee represented a small step in the right direction. Many of those involved went on to assume leading roles in Weston conservation, and eventually something very much like what we had proposed did in fact take place, only by a different route. Those who intend to have community farms should understand that it may take more than a few false starts. Perseverance is by far the most important skill you will need.

At the time, however, we seemed to have exhausted all the obvious possibilities for expanding Green Power's fifteen acres of projects to the rest of Weston's conservation land. The easiest course for the town was to let the land lie in a state of nature, which cost nothing and caused little controversy. I felt discouraged. Brad and Sonia split for Amherst to finish their schooling at the University of Massachusetts, where Sonia now works as an entomologist specializing in integrated pest management. After three years in Weston, I moved a few miles away to the Concord woods near Walden Pond and for a year or two was content to live deliberately, fronting only the essential facts of life. I quit working for Green Power in the spring of 1979 and earned what little cash I needed as a casual private contractor, operating mostly in Weston, working in woodlots, and trying a little carpentry and other odd jobs. I did learn by this experience that I was unlikely to succeed as a landscaper, housepainter, or handyman, but that I liked chainsaws more than I expected.

While I was living in Concord, Doug Henderson and I kept conspiring about how to implement our ideas for Weston's conservation land. It wasn't just a matter of expanding Green Power—we believed that when Bill McElwain retired Green Power would be expunged. There were influential people in town just itching to write that misadventure in flagrant socialism out of the budget. We decided we had to organize a group outside of town government that could work effectively with the town, but we weren't sure how to go about it. Finally, we met with board members of Codman Community Farms in neighboring Lincoln, a nonprofit organization founded in 1973, and here found our model. As a membership organization we could raise capital through tax-deductible contributions from those in Weston who supported our goals, building a political constituency at the same time. We could put on events to raise money and spread the message and sponsor educational projects. Most important of all, we could contract with the town to manage conservation land.

A nonprofit organization would relieve the town of the burden of establishing a full-time staff position and would allow us to start with small projects on a part-time basis. We would be an independent organization controlling our own purse strings while at the same time working closely with the Conservation Commission on a comprehensive program for public lands. This way the commission could set policy and oversee the care of town land, its appointed role, without becoming bogged down in the day-to-day business of managing it. As a nonprofit, we would be free to carry out similar projects on private land to bring in extra money or otherwise advance our goals, as long as these projects fell within our educational purposes and did not bring financial reward to any individual, beyond the wages paid to staff members. This was exactly the flexible approach we needed.

Doug set about assembling a board of Westonites to form such a corporation. We lifted Codman Farms' bylaws almost verbatim. One of our cofounders, Martha Gogel, found a lawyer willing to do pro bono legal work to get us incorporated. We took the name Land's Sake because Doug's grandmother had been fond of that expression and because we were acting for the sake of the land. So in the summer of 1980 Land's Sake was born. I moved back to Weston and joined the board in the fall. We sent out a townwide mailing announcing our birth, explaining our mission, and asking Weston residents to support us as dues-paying members. We were so green we neglected to put our names on the mailing, so that many people had no idea who this mysterious Land's Sake group was. Still, we signed up a few dozen charter members—including such prominent conservationists as Bill Elliston, who supported Land's Sake strongly in the last years of his life (Harriet Elliston still sits on our board). It was a small but decent start. Today, Land's Sake boasts of more than four hundred members, and they contribute some thirty thousand dollars every year.

From the beginning Land's Sake had a firm grasp of its three central principles: provide ecologically sound care of land in Weston, involve the community and particularly young people with the land, and make the program pay its own way as far as possible through the sale of products and services. I spoke of these principles as a three-legged stool, which Doug always said made it sound like somebody was being milked. The first, our primary reason for being, was a pointed departure from the prevailing preservationist philosophy toward open space. Suburban conservation land was not wilderness, we argued, but a diverse mixture of woodlots and open fields. Rural character required rural care, which meant active farming and forestry. We aimed for an attractive balance whereby the land, as we put it in our mailing, would "benefit from our presence, rather than need to be protected from us."

Our second principle was that whether or not the land needed our care, we needed to care for it. The community would benefit by being involved with the land. We especially wanted to impress this habit on young people, in the manner of Green Power, to enliven the experience of growing up in Weston. Finally, we wanted Land's Sake to be as self-supporting as possible. Earning a good part of our income would free us from the constraints of town budgets and private fund-raising. As some of our board members pointed out, the educational benefits would be more meaningful to the children involved if the projects were legitimate economic activities, not just make-work. From the beginning, the challenge for Land's Sake has been keeping these three ecological, educational, and economic legs sufficiently even so that our milking stool stays flat enough to sit on.

By working closely with the Conservation Commission we hoped to develop an integrated program appropriate to all of Weston's diverse conservation land, from open fields to dense forests. Such a program would find valuable roles for young people and produce high-quality natural products for local sale. It would strike a balance between protecting natural ecosystems and making sustainable, productive use of the land. It would offer ample educational and recreational opportunities, and it would bathe the community in attractive rural character until it glowed. That was what Land's Sake set out to do in 1980, and that is what we are still working toward almost two decades later.

One of the first jobs contracted to Land's Sake was tending Weston's trail system. During the years that Weston was acquiring its conservation land the Forest and Trail Association worked to keep existing trails clear and to add new ones. Local equestrians wanted to stay in the habit of riding from one end of town to the other across what had been their friends' estates, while many suburban newcomers avidly pursued birdwatching and walking in the woods. By the 1970s these traditional trail users had been joined by joggers and cross-country skiers, and by the 1990s the mountain bikers had arrived as well. Today, all of this so-called passive recreation coexists with only occasional conflicts (horses can ruin ski tracks, for example, and the mountain bikes can ruin everything), sharing sixty-five miles of trails.

Among the many people who have worked to maintain Weston's trails, one person stands out. In his positions as treasurer of the Forest and Trail Association (where he succeeded Bill Elliston) and chairman of the Conservation Commission (where he succeeded Julie Hyde), George Bates has seen to it that trails are well kept and well marked, fire roads and causeways are graveled and graded, gates and culverts are installed, fields are mowed, overlooks are kept clear, and dams and bridges do not wash away or are put back where they belong if they do. At first much of the routine trail work was done by volunteers, but in time this proved inadequate (as it nearly always does). Recognizing this, George hired Land's Sake to look after the trails for the Forest and Trail Association, who in turn looked after them on behalf of the town. The arrangement supplied Land's Sake with a reliable income and gave the town efficient means of caring for its property.

Save for poison ivy, mosquitoes, and the odd murderous goshawk, the work of maintaining the trails was serene and pleasant. We clipped encroaching brush, cleared away windfalls and widowmakers, cleaned up after storms, built footbridges, rerouted trails around irredeemable quagmires, laid out new trails, and put up trail markers. Caring for the trails gave us the chance to systematically explore all of Weston's community-owned land. After a few years there was scarcely a corner of the entire two-thousand-plus acres that I hadn't walked through at one season or another, thinking up possible projects and discussing them with my eager young coworkers. Thus we generated a range of ideas for almost every acre of public land in town. Only a few of these have been realized, but, as you will see, those few are enough to fill a book. Our philosophy at Land's Sake was not that every inch of conservation land should be managed according to some grand preconceived plan, but rather that we should be ready with good ideas when opportunities arose—ideas that formed a diverse but consistent pattern. These were modified first through discussions with the Conservation Commission and other interested parties, and then again as we put them into practice. In just this organic way, an active management program for Weston's conservation land continues to take shape.

A good trail system is fundamental to a livable community. Walking Weston's trails enabled me to see the town whole, at the pace for which it was designed. Weston's open land and trails and its charming center and the winding roads radiating from it were just ghosts of what had been an ideal pattern of rural settlement, scaled to going afoot. I discovered that I had passed beyond merely being able to discern that pattern—I wanted it back. That would be impossible in Weston during my lifetime, of course, but it helped organize my thinking about what we were doing. It might be possible, out beyond Weston where towns were just beginning to undergo development, to make better modern versions of a working New England town. Our first steps could help provide the model.

Is this aspiration just syrupy nostalgia for a bygone agrarian age? Every age has its golden moments, its dark despair. What I wanted to restore as far as possible was not the admittedly imperfect flesh of the past, but something of the physical skeleton, the hard bones of it. The automotive-driven residential and commercial sprawl of suburbia is monstrous compared to the human-scale order that once prevailed. The expanding conurbation of beltways, office parks, strip malls, endless housing tracts, and ceaseless traffic should be halted and gradually replaced with something better, for the sake of both sustainability and sanity. We should advance to the past, to a model of settlement very much like the preindustrial New England town. I am completely serious. It is within our power.

The classic New England town contained a central village surrounded by farms and woodlots out to a distance of three or four miles. That was a distance that a person on foot or driving an oxcart could cover in an hour (it was quite similar to the travel time people will tolerate to pursue their daily business today). Five miles across with a central village and abundant open space is a community structure that still makes eminent sense, in spite of the automobile. We are not going to see the return of the oxcart or buckboard, but for much travel within town the bicycle works very well over just the same range as the horse. Walking and cycling are marvelous ways to get around, if the roads are reasonably safe and pass through pleasant countryside or there are off-road trails. Our culture is not about to junk the car—I have given up hoping for that—but we could certainly make more limited use of it within our homeplaces and be all the healthier for moving under our own power. This outcome is still achievable, to varying degrees, in rural areas that are turning into suburbs.

It wouldn't take that great a modification of the way New England towns are now being resettled to create modern versions of this sensible form. Our towns could be tied to the metropolis by fiber optic lines and reliable commuter rails, reducing through traffic. Within these towns, a dense central village surrounded by smaller outlying clusters of residential neighborhoods embedded in a landscape of active farmland and extensive forest would invite walking and cycling. The result: more livable communities than the suburbs we now endure, with their harried car traffic, constant racket and haze of exhaust, and veil of dispersed homes and yards drawn over the face of the land. Those who protest that no one has the time to live this way are merely making my point for me—if we don't have time to live well, what have we achieved with all our economic growth? At least Thoreau's farming neighbors, who were suffering through the commercialization of their agrarian world and going bankrupt, led lives of *quiet* desperation. The desperation of their commuting descendants, trapped in their speeding cars, is frantic and dangerous.

Obviously this vision arises from an idealization of the New England town. As a historian I am well aware of flaws in the region's past. Those picturesque country roads may have been a muddy morass, and many of the outlying farms they passed may have been foul to look upon, hardscrabble and unkempt. The countryside was much abused a century and a half ago, stripped of trees and thick with scrub. Filthy milltowns, denuded forests, and polluted streams were erased from the charming vistas painted by the Hudson River school. A close reading of Hawthorne, Melville, and Frost reminds us of the grim side of pastoral New England; Henry Thoreau died young of tuberculosis, and his brother John died even younger of lockjaw, scratched by his own razor. Many small-town inhabitants were as narrow and mean-spirited as the most self-absorbed suburbanite today and perhaps less apt to leave you alone. Agrarian life certainly had its flaws—some of them intrinsic to a society less technologically advanced, some of them brought on by the same market forces that drive us to ruin today.

Yet none of the agrarian past's real flaws detracts from the essential attractiveness and utility of the New England town *form.* I am claiming not that small rural towns ever were or ever could be utopias, but simply that the structure is better suited to ecological and social responsibility than the present runaway suburbs of detached individuals who spend what seems like half their waking hours in their cars. And the place would certainly look better. I know these homely ideas are not new—and I think I understand the self-reinforcing drives of eternally unsatisfied consumption and frustrated individualism that keep them from being adopted. That Americans continue to vote with their feet for the accelerator over the footpath only tells us we need to redouble our efforts to persuade our fellow citizens, through working examples, that there are more livable possibilities. Speeding after rural character in consumable private pieces destroys it. Suburbia is a good urge gone wrong. There is be a better way to get back to the countryside and to small communities than this.

Weston, it seemed to me in 1980, offered a taste of a better possibility by setting aside almost one-quarter of its land as open space. Towns like Weston and Lincoln hadn't retained nearly as much farmland and forest as I might have liked, but they had achieved something far better than total "build out." They afforded a strong sense of what could be achieved by other towns employing some of the same tactics earlier in the game and taking them another few steps. These suburban towns demonstrated that to keep an appreciable part of the landscape open, a suburbanizing community had to acquire control over it. But that was only the first step. To build support for protecting more land, we needed to encourage active community involvement with it. We needed to see to it that the land was not only walked, but worked, so that it would become a vital part of community life. Open space could not compete with residential development on the appeal of window dressing and wildlife habitat alone. The land had to become not something peripheral, but an integral part of the way the community defined itself. That is where groups like Land's Sake would come in.

Protecting conservation land and putting it to use amounted to the creation of a new kind of *commons* system. In a traditional commons system some portion of the land was jointly owned and managed, and citizens of the community held specified rights in that common land. Traditionally, of course, this meant productive economic engagement—that is, the land was used. To have a real commons was something more than enjoying a park, in other words, more than tossing a frisbee on the town green. Weston had the common land, but what about the engagement? Even though the town wasn't about to allow every householder to cut a cord of wood and pasture two beasts annually on its conservation land, what Land's Sake proposed through community farming and forestry might be the modern equivalent. We did not advocate that the land be managed strictly by professionals, like a national forest. We proposed to employ foresters and farmers, but they were to answer to a local board of citizens, the Conservation Commission. They were to employ local young people, and they were to produce food and firewood for local consumption. This struck me as constituting a modern commons, in the full sense of the word.

Would this idea of common land that people actually used strike a chord in a town like Weston? Interestingly, common landownership had played an important but largely forgotten part in New England history. Many of the first New England towns (including Watertown, from which Weston later sprang) began with commons systems brought over from England. But those commons had faded away three centuries before. It wasn't clear how that ancient history might be connected to what was happening in the late twentieth century. Still, it was a nice precedent, an idea worth exploring. I was learning that communities such as Weston had a way of remembering things subconsciously for a long time.

It *was* clear that the unbroken tradition of the town as a bounded place with a democratic, voluntary form of self-government was a great help in this revival of the commons. Weston was a place that was used to thinking of itself as a community, where people knew from long practice how to work for the common good in spite of their differences. True, only a minority of the residents participated actively in town affairs, but it wasn't just people who had lived in Weston for generations that upheld this tradition. Just as often it was newcomers to town, whether from Boston or Houston, who became involved in town affairs. The institutions of local self-government had endured in spite of suburbanization and geographic mobility, pulling new citizens into the practice of civil self-government as they arrived. This was encouraging.

From this experience of caring for the community as a whole came a willingness and an ability to assert a common interest in the care of land, even in a region with as strong a belief in the rights of private property as any other. Citizens in towns like Weston had seen the need to protect open space for the good of the community and were able to act on it by imposing local taxes upon themselves in order to buy private land and make it public land. Conservationists had persuaded a majority of voting townspeople that the town should once again own some land in common, three hundred years after selling it off. The task before Land's Sake was to demonstrate the full ecological, economic, educational, and esthetic value of this inchoate commons, which meant involving the community more actively with the land. But at Land's Sake we controlled no land and spoke for nobody but ourselves. Where would we begin?

Doug Henderson bringing blueberries and flowers from his garden to the Case farm-stand. (Neil Baumgarten)

Chapter 2

Market Gardening

When we formed Land's Sake in 1980, we had very little capital, almost no equipment, and no commitment from the Conservation Commission to put any of what we contemplated into practice. We inherited a few projects that I had begun as a private contractor: maintaining trails for the Forest and Trail Association, planting new apple trees in the orchard on Concord Road, cutting firewood for private landowners. We had hopes that more projects along these lines would attract support and that they could provide interesting, educational work for young people. In our dreams were dozens of possibilities, a potential pattern of community involvement with the land that radiated from gardens to hayfields and pastures, on through orchards and maple groves, and out to the forest that covered the bulk of Weston's land. We could vaguely see what we wanted to accomplish on these largely idle town lands; now we had to make our way toward it. Slowly, over the next decade, working closely with the Conservation Commission, we succeeded in making a surprisingly large part of that vision real. Our greatest success was the market garden at the Case Forty Acre field, the place most people now identify simply as Land's Sake.

The Case Farm was born more or less by accident. When Land's Sake sent around that first townwide mailing in the summer of 1980, one person who read it with interest was Peter Ashton, the newly appointed director of the Arnold Arboretum at Harvard University. Ashton, a renowned tropical ecologist, had just moved into a farmhouse at the Case Estates in Weston. The Case Estates had been left to the arboretum in the 1940s by Louisa and Marian Case, the last of a wealthy Weston family. The estate became a suburban field station supporting the arboretum's main grounds in the Jamaica Plain section of Boston. Three parcels, comprising a little more than one hundred acres, lay along Wellesley, Ash, and Newton streets about half a mile south of the center. The property included high, rolling glacial till to the south and dropped northeast across Newton Street to nearly level glacial outwash on the parcel called the Forty Acre field. The Case Estates presented a delightful mixture of woodlots, specimen trees, orchards, azaleas, rhododendrons, perennial flowers, nursery beds, and elegant farm buildings left over from the time of Marian Case. The land was Weston's garden spot, the town's pride and joy.

By 1980, however, the arboretum's new director was in a bind: his budget was tight, and the Case Estates began to appear peripheral to the arboretum's mission. The Forty Acre field, in particular, had no active collections and was essentially sitting idle. Its plantings of crabapples, cherries, and other flowering trees scattered in clusters among old hayfields were certainly beautiful, but of no great horticultural significance. It was not easy for Harvard to tolerate an impecunious offspring sitting on unused land in a wealthy town like Weston, especially as the real estate market began to froth and bubble in the early eighties. Under pressure to sell the Forty Acre field to help float the rest of his operation, Ashton thought perhaps some part of the parcel could be used to grow fruits and vegetables, giving the land an active program and perhaps even generating a little income. He contacted Bill McElwain to inquire whether Green Power wanted to use the land, but Bill had his hands full. About this time, Land's Sake's manifesto advocating productive use of Weston's neglected open space dropped into Peter Ashton's waiting mailbox. If you declare to the world that you are going to do something, you never know who may take you up on it.

Doug Henderson and I went to meet with Ashton and his staff on a brisk day early in 1981. We walked over from Doug's house through an intervening block of conservation land; there were some interesting old earthworks and drainage ditches at the back of the Case Estates that Doug wanted to show me. I think the idea of walking out of the snow-filled woods and arriving at the arboretum's *back* door appealed to our sense of what Land's Sake was and of how Weston had been and should again become: we were neighbors coming over to visit and to do a little cordial business by and by. We proposed growing strawberries and raspberries, using butternut squash to prepare the ground. We would hire kids to do the cultivation in the manner of Green Power, but older kids under a tighter commercial rein. The arboretum agreed to fund the project if we could come up with a convincing business plan. We said we could, but in fact, we had only the haziest idea of what we were getting into.

Doug and I stepped back out into the wind and snow highly elated: it suddenly appeared as if the Land's Sake seed we had planted was actually going to blossom with the spring. I did a quick search of my files for information on strawberries and raspberries (which I had never seriously grown) and knocked out a five-year farm plan with conservative income projections that still showed a one-thousand-dollar-per-acre profit at the end of the day. The arboretum accepted it. We could scarcely believe our luck: Harvard University had become our sugar daddy.

That spring we broke up a few acres of old sod in two long strips between the specimen trees on the Forty Acre field and got down to tilling the soil. We recruited a few high school kids and planted a quarter acre each of strawberries and Heritage fall-bearing raspberries and an acre of butternut squash with a few hills of Ambrosia hybrid cantaloupes thrown in for good measure. We cultivated the berries as infrequently as possible, staying within our budget. In the course of the summer we had Codman Farm in Lincoln come over and plow up a few more fields among the trees. I borrowed Green

Power's old Farmall tractor and periodically harrowed the rough furrows down with the Case Estate's disk, putting under a succession of exuberant buckwheat crops for green manure. We were off and farming.

I can't exaggerate the joy I felt farming the Case land those first few years. It was like what I had felt at Green Power half a decade earlier, only this was my farm, my contribution to the world. Of course, it was not my land, but that didn't matter, and it still doesn't. I came to care for it in a way that has little to do with legal ownership or exclusive possession. I had a vision for what that land could become that was destined to come true after a long, hard struggle. In that sense the land is mine, as much as land can ever be anyone's.

I was not alone: Doug's steady hand and Martha Gogel's indomitable energy were behind me. Without them, Land's Sake would not have survived its early years. All that spring Martha tooled around Weston in her old Volvo station wagon hauling Doug's Troy-built rototiller behind her in the trailer Doug had built, which along with my chainsaw and pickup and Doug's Garden Way cart made up Land's Sake's entire stock of farm equipment. Martha tirelessly rototilled gardens for new Land's Sake members. Doug rototilled his share of home gardens, too—I always smile when I think of the former U.S. ambassador to Bolivia out tilling gardens for twenty-five bucks a pop. Their work brought in a little cash to keep the organization afloat while I was finishing our first firewood cutting contract in Jericho Forest and getting the Case land planted.

Those were inspiring times, when the farm was young and growing fast. What we were doing was a suburban miracle. An exceptionally beautiful piece of land that had been slumbering for decades was springing to new life, with fruits and vegetables growing among flowering trees from every temperate corner of the planet. I remember Bill McElwain coming over one lovely May morning to till up some land for squash. "Who would believe this is going on just ten miles from downtown Boston, eh, lad?" Bill asked, gesturing grandly from the tractor seat with his cigar. Of course, without Bill's pioneering at Green Power, it would never have happened at all. If our plans succeeded, what Bill had started would not be lost.

The first years of Land's Sake were exhilarating, but they were also a hard lesson in what it takes to grow fruits and vegetables on even a small commercial scale. In spite of my seasons at Green Power and Martha and Doug's lifelong gardening experience, we had a lot to learn about growing crops by the acre under a tight budget, let alone selling what we grew at a decent profit. We had some differences about how best to proceed. When we were ready to disk the field and plant butternut squash, Martha came by and announced that we could get as many loads of manure as we wanted from friends who needed to clean out their stable. I told Martha I didn't think a few loads of horse manure, which in Weston was mostly wood chips anyway, would make that much difference spread over an acre of land. Besides, we had no efficient way to spread it. Martha said we should at least put a little manure in the hills and get the squash off to a good start. I said we couldn't justify the labor cost. Martha played her favorite trump card, which was that her labor didn't cost Land's Sake anything, and she would do it herself. Fine.

When I arrived with the Green Power tractor the next day, I discovered that Martha had been hauling manure nonstop. The entire field was covered with piles of road apples and sawdust, too big to disk through. Her friends had a little front-end loader to fill her trailer—I hadn't counted on that. So I wove the disk among the piles, fuming, and then Martha and the kids and I carted a little manure to each squash hill as we planted the seeds in them. True to her word, Martha spread the rest of the manure over the next few weeks. Later that summer, once the vines ran and the disk no longer fit between them, she guided the rototiller among the hills, in her bare feet, battling the immortal witchgrass as long as she had daylight to steer. Such inefficiency overcome by heroism typifies community farming.

The butternut squash bore a bumper crop, which I knew very well was because the land had lain fallow under grass for half a century rebuilding organic matter, and Martha knew was because of her horse manure. The Ambrosia cantaloupes did spectacularly well. The berries we had bravely planted into fresh-tilled sod were mediocre at best, but we planted more into cleaner ground the next spring. Over the next four years we steadily expanded our plantings until some ten acres were under cultivation. About half was in raspberries and strawberries. The rest was planted in a wide variety of vegetables, in rotation with the strawberries and green manure crops. Most of the hoeing and weeding was indeed being done by hard-working high school students, and we were building support in the community for our efforts. We felt sure we were on the road to success.

The only hitch was, we weren't making Harvard any money. In spite of all our hard work, the profits I had projected failed to materialize, and our sugar daddy was growing less indulgent. There were a few reasons for this failure. One of them was witchgrass (*Agropyron repens*), the stoloniferous scourge of the Northeast, which choked our raspberries and strangled our poor strawberries in their beds. I had never grown crops like these at Green Power, and so learned the hard way the dangers of planting perennial berries into old sod land. We were on the right track in trying to clean the ground the year before putting in the berries, but we should have been planting pumpkins, which are bigger and meaner than butternut squash. We weren't yet fanatical enough about keeping our plantings clean. We were employing labor frugally, which is a good instinct as long as you know what work absolutely must be done. The strawberries would look all right when we mulched them for the winter, only a little weedy. By spring, the plants had all but disappeared under a sea of sheep sorrel and witchgrass, yielding only a miserly crop of small, sour berries. The arboretum urged us to use herbicide to control the grass, but we refused. We were using chemical fertilizer in moderation to ensure decent crop yields while we built up the soil with green manures, but we did not intend to apply chemical herbicides and pesticides to food. Most of the vegetables did fine under this regime, but the berries were supposed to be our main cash crop, and they were hurting.

Even worse, we didn't yet have an adequate retail outlet for the produce we were harvesting. The arboretum had ruled out selling through a farmstand on the property, the obvious solution. We assumed this was because of Harvard's liability worries, but they later told us that they had assumed the town would not welcome a farmstand. That was an unfortunate misunderstanding—under state law the town could not have stopped us from selling our produce on site even had they wished to. We peddled some fruits and vegetables to local restaurants, roadside stands, and whole food stores. The chefs at the two country clubs in Weston paid well, but they couldn't absorb enough of our harvest. We set up a little stand on the sidewalk in Weston

Center, and Martha hauled produce to various farmers' markets in Boston. But we just weren't moving the necessary volume.

The butternut bonanza we stored in the Case Estates barn, then carefully graded and repacked it in standard wire crates and hauled truckloads weekly to the Northeast Food Coop warehouse in Cambridge. It was satisfying to deal with the coops, but selling squash wholesale is a hard, hard way for small farmers to make a living. The second year Martha managed to swing a deal to supply the Harvard dining halls, so undergraduates could eat squash from their own land, which I'm sure made a deep impression upon them. The catch was, we had to clean and peel it before the Harvard kitchen would accept it. So we'd get together for peeling parties one evening a week in someone's kitchen and go wild peeling, cubing, and bagging squash. This was fun once or twice but financially a bust.

After four seasons in the red, Land's Sake exhausted the arboretum's patience. Income did not appear to be trending upward, nor expenses down. Early in 1985 they informed us they had decided to lease the farm we had created to a commercial grower from a nearby town. That day was one of the darkest of my life. Several members of our board thought perhaps losing the farm was for the best and would give us a chance to concentrate on other projects on town land, but Doug and Martha and I knew how crucial the Case Farm had become to Land's Sake, and we were determined to get it back. I went and begged. The arboretum relented enough to allow us to grow flowers and other specialty crops on the front acre of the property, as long as we did not compete directly with the new farmer. We swallowed our pride and took what we could get, the most difficult (and also the best) decision we ever made. We didn't yet know that our friends at the arboretum were in the midst of losing a much larger battle over the fate of that land. Later that spring, Harvard announced that the Forty Acre field would be put on the market. All hell broke loose in Weston, and Land's Sake was smack in the middle of it.

The Rise and Fall of Market Gardening in New England

Was a small, labor-intensive market garden employing young people in the suburbs an economically viable proposition? Most people would have told you it wasn't. In 1985 the arboretum staff informed Land's Sake that much as they admired our spirit, we had proven abundantly that trying to farm ten acres with rototillers and teenagers just couldn't cut it. Besides perversely turning our backs on mechanization in favor of unskilled hand labor whenever possible, we also abstained from using most agricultural chemicals and later went completely organic. How could we possibly compete with the productivity of modern industrial agriculture? With California's Central Valley pumping produce across the land, maybe farming no longer had a place in Weston, like everyone said. "I've lost a lot of faith in Land's Sake," said one arboretum horticulturalist to me with genuine regret that winter, and although I hadn't lost mine, I had to think hard about what we were doing. Maybe it really wasn't possible to be ecologically responsible, educational, and profitable all at once. What was the missing link that would make all these things work together? Small farms run by highly skilled growers had been going steadily out of business in our area for generations, until only a handful were left. What made us think we could succeed with a bunch of schoolkids? Why was it so important that we succeed?

Land's Sake sees itself as part of a movement to create a sustainable local food system; an effort to revive farming in New England and to convince consumers that eating close to home is healthier for both the land and themselves. Today, Massachusetts imports from out of state more than 90 percent of its food and nearly 80 percent of its fresh fruits and vegetables.[1] Much of that food is produced at great ecological cost, and much of it is denatured and loses its vitality and flavor as it makes its strange, convoluted journey from the earth to our mouths. Certainly it would be better if we could grow food closer to home in more sustainable ways and eat it pure and fresh. But how much of the region's food did New England farmers ever produce? What is this land fit to grow, and why did we stop growing it here? These are questions that need to be answered, not with blind enthusiasm, but with a careful look at New England's farming history as a sequence of complex ecological systems that gave way one to the next for powerful economic reasons. The modern, fully industrialized food system that today provides for us almost entirely from afar began to emerge after World War I and swept Massachusetts all but clean of farms after World War II. Before that, New

Englanders did grow much more (though by no means all) of their own food, particularly milk, eggs, fruits, and vegetables. We need to begin by examining this bygone food system, the reasons for its demise, and the shortcomings of the industrial food system that has replaced it.

A century ago, Boston was surrounded by thousands of market gardens supplying fresh produce in season to customers in the city and its satellite factory towns. Weston lies in Middlesex County, the largest and once the most agriculturally important of the counties adjoining Boston. In 1910, there were well over ten thousand acres producing vegetables in Middlesex County alone. By the 1990s that had shrunk to little more than two thousand acres. Weston, a hilly town lacking as much prime tillage land as some of its neighbors, still reported nearly three hundred acres in commercial fruit and vegetable production in 1905. Today, Weston has (to my knowledge) only twenty-five acres left in commercial fruit and vegetables—plus a few thousand square feet under vinyl in two greenhouses belonging to Julie and Peter Hyde, who grow hydroponic tomatoes. Farmers in the suburbs of Boston once excelled at growing such crops as asparagus, strawberries, cucumbers, and rhubarb. By the turn of the century they had pioneered the use of coal-heated greenhouses to extend the seasons of their crops, and many were growing flowers as well. Fruits and vegetables were once grown here in the suburbs and grown in plenty. The story was the same around every American city.[2]

Market gardening was part of a commercial agricultural system that took shape in Massachusetts about the middle of the nineteenth century and flourished until shortly after World War I. Even in their heyday, of course, market gardens around Boston never comprised a completely sustainable local food system—although they did supply a much larger measure of food within the region than we grow today. From the beginning, local truck farms were only a small part of an emerging national system of extractive industrial agriculture. By 1850, the establishment of a national rail network meant that farmers in Massachusetts faced withering competition from newly established midwestern farms in the production of traditional staples like grain, meat, and wool. But this did not lead to the collapse of New England agriculture, as is widely believed. Quite the contrary: New England farmers made use of the influx of cheap feed grain to specialize in more valuable crops, meeting burgeoning urban demand within the region.

The switch to more specialized farming was very successful in economic terms. The gross value of agricultural production in Massachusetts increased rapidly in the late nineteenth century, peaking during the first decades of the twentieth century. Dairy farming (see chapter 3) was far and away the leading sector at the outset of this new, thoroughly commercial agricultural system. Poultry also made money, increasingly by the conversion of purchased feed into eggs and meat in commercial chicken houses much larger than traditional farmyard flocks. Hay remained a profitable cash crop on many farms, as horses remained the leading means of local transportation until about World War I. But fruits and vegetables grew most rapidly of all, catching up to dairy products in value by 1920, before going into equally rapid decline.

The period from 1850 to 1920 was what contemporary agricultural leaders called the era of concentrated products in New England farming. The emergent farm system relied heavily on imported feed grain, but also made very effective use of local farmland and organic wastes to produce good food. High-value products such as milk, eggs, and fruit were called concentrated to distinguish them from lower-value bulk commodities like the grain being grown on cheaper land in the Midwest. These products were often a concentration of that grain itself into something more valuable. It is a great shibboleth of New England history that after the railroad trunk line went through in 1850, cheap midwestern grain undercut New England agriculture. This oversimple lesson of history is mostly wrong. Farmers in New England actually bought this cheap grain in great quantities, fed it to their cows and chickens, and turned it into far more valuable milk and eggs. They concentrated it. As a bonus, in the process they received an augmented supply of manure from the extra nutrients flowing onto their farms.[3]

Finding enough fertilizer had been the conundrum of New England farming for more than two centuries. Traditionally, nutrients had flowed from the hayfields, through the livestock, and onto the tilled land. Imported grain reversed this flow and produced surplus manure: that is, more nutrients were being generated by the Massachusetts cow than were being demanded of the Massachusetts soil to feed her. This extra manure could go one of two ways, depending on the farm: to fertilize either truck crops or corn fodder and hay—part of which came back to help feed the cow and part of which could be sold. To actually have enough dung to dress the mowing land and to be able to sell hay without selling the farm in the process was something new and exciting in New England agriculture. And it was made possible largely by the massive through-put of cheap feed grain.

Hay that left the farm was sold primarily to feed horses, whereby it joined another stream of feed grain coming from the Midwest, this one made up mostly of oats. This was the great era of rail transportation, but steam locomotives did not replace horses. The trains created more work for horses in an expanding economy, and they brought sacks and sacks of oats from the heartland to feed them. Horses were still needed for short hauls, and there were far more short hauls of both people and freight to be made. The manure from city stables was shipped out to suburban market gardens, another source of recycled nutrients for fertilizer. Farmers also bought large loads of spent malt from local breweries. Into the early twentieth century, market gardens surrounding America's cities served not only as an abundant source of fresh local produce, but also as a partial recycler of vast quantities of urban organic wastes.

The system was by no means completely circular or environmentally praiseworthy on all counts. In the first place, a large percentage of these locally recycled organic nutrients, if traced back through the cows and horses and mash kettles to their sources, would have been found to originate in the soils of the Midwest, which were being mined by ruinous extractive grain farming in what one historian has called "the most stupendous bargain counter in the history of agriculture."[4] Second, although there was great interest in the recycling of human waste (night soil, as it was called) during the nineteenth century and many experiments were tried, no economical method of returning these nutrients to the soil on a large scale was ever developed. By the late nineteenth century, many market gardeners (and other eastern farmers) were supplementing their organic manures with purchased commercial fertilizers. Most of the nitrogen in these mixtures came from various organic sources, but phosphate and potash came increasingly from mines, just as they do today. Finally, highly toxic pesticides were already being used by the 1860s, chiefly a variety of deadly compounds that employed arsenic to kill insects. Turn-of-the-century market gardening was hardly organic farming.[5]

The era of concentrated products was far from being a regionally self-contained, organic food system, but it did have the virtue of bringing together suburban land and urban wastes to produce a significant portion of our food. A variety of local and imported sources for human food, animal feed, and farm fertilizers were effectively combined. Such was the agricultural system prevalent in the great era of expansion of American cities and heavy industry, the age of iron and coal, streetcars and railroads.

A well-run small farm on the urban fringe in turn-of-the-century Massachusetts functioned as both an intensive supplier of food and a recycler of nutrients. One thirty-acre operation described in the state agricultural report for 1888 will serve as an example. Two or three acres of cropland were laid down to grass and clover every year and left in hay for two or three years—so about one-quarter of the farm was always in grass and legumes. These six to eight acres provided hay for five workhorses and five dairy cows, with some surplus to sell to city stables. The livestock doubtless consumed purchased feed as well, since no grain production is mentioned in the report. The abundant manure from these animals was applied to the rest of the farm, which produced vegetables and fruits, the main cash crops, for market. Back from the city came the manure from one hundred horses, which was spread on the land along with lime, ashes, and a few tons of concentrated fertilizer like nitrate of soda, muriate of potash, and superphosphate. Farmers liked to use animal manure along with agrochemicals because they recognized that it improved the tilth of the soil even as it added nutrients. Some market gardeners applied as much as forty tons of stable manure per acre! Truck farms farther from the city had to rely more heavily on plowing down green manures like crimson clover to build humus. Now here was an exemplary use of suburban farmland that only a severe chemical imbalance in the modern brain could have caused us to forget.[6]

The sensible regional food system of a century ago unraveled in the onrushing age of oil, chemistry, electricity, and the automobile. After 1920, suburban market gardens were indeed undercut by western competition, and then the very land was cut from beneath them as well. Refrigeration and rapid transportation made it possible to ship fresh produce year-round from the West Coast and Florida, and after World War II frozen vegetables also gained a larger share of the produce market. The growing of enough fresh produce to keep millions of crispers and freezers filled year-round required major new technological heavy lifting in agriculture. From the Florida Everglades to Washington's Columbia Valley, natural hydrology was reorganized on a massive scale through irrigation and drainage projects. The cultivating of fruits and vegetables in California's Central and Imperial valleys and in irrigated parts of Texas, Arizona, Idaho, and Washington required damming of several major rivers and ultimately the diversion of a large part of all the water that flowed in the West. This was accomplished with a correspondingly massive influx of public money, beginning with the Reclamations Act of 1902. Those of us still growing vegetables in the East today have not forgotten that the so-called free market competition that ruined most eastern growers was largely funded by American taxpayers, who then and now supply western agribusiness with subsidized water.[7]

The reorganization of nature to support industrial farming was extremely energy intensive. It required unprecedented amounts of earth-moving engines and earth-replacing concrete—it all ultimately rested, and still rests, on fossil fuel. Food produced in this way will remain economical only so long as energy remains cheap. Cultivation on such a scale was beyond the reach of animal muscles and organic manures; it required gas-powered tractors and inexpensive petrochemical-derived fertilizers to really take off. It was possible to grow hundreds of acres of the same vegetable in one place only with the help of powerful new pesticides, and harvesting those vast acres of produce took armies of underpaid migrant farmworkers or new machines. Market gardeners in the suburbs of Boston were up against a new capital-intensive, highly mechanized mode of vegetable production that flowed from the oil bonanza of this century and from novel scientific abilities to resynthesize nature at the scale of both hydro dams and carbon bonds. Generation after generation, local growers were worn down by this competition until they all but disappeared.

Smaller growers in the Boston suburbs also took up the new technologies and scientific methods, of course. They, too, turned increasingly to tractors in place of horses, to chemical fertilizers in place of manure, and to chemical pesticides. But the most profound impact of the age of oil on suburban farming was the steady rise in the value of land that resulted from spreading residential development. Before the automobile, suburban living for the middle class had been mostly restricted to the range of streetcars—only a few wealthy industrialists commuted by train from homes as far out as Weston. With the automobile came a broad new wave of expansion that is still rolling outward today. This residential tide steadily inundated suburban farmland. Farmers faced both the rising cost of land taxed at its much higher residential value and the undertow of temptation to give up a demanding way of life and cash in the farm for development. After 1920, many a Massachusetts farmer told his children as they came of age that there was no future in farming and urged them to get into more profitable lines of work. A generation or two earlier it had been possible to switch from general farming into dairy or market gardening and make a decent living; now it seemed there was nothing left for suburban land to grow but houses.

Land of all kinds was developed, but flat, well-drained tillage land was cheapest to build on, accommodated septic systems easily, and was often the first to go under. Land that had been growing fruits and vegetables was steadily settled upon by people who drove to supermarkets to buy produce trucked in from the far corners of the earth. The process is still under way as the suburbs wash further out into the hinterlands, beyond the beltways that have spawned new commuter riptides of their own. Today, the last of the farmland best situated to supply metropolitan Boston with fresh produce is fast disappearing.

Why should we miss this local farmland? First, because it enriches our lives: it is beautiful, it makes healthful outdoor work available for our young people, and it produces better food than we can otherwise obtain. But there is another reason. The now-perfected system of industrial food production and consumption, stretching from suburban sprawl at one end to a sprawling transcontinental supply network at the other, makes only illusory economic sense. As long as energy is cheap, ecological costs are mortgaged against the future, and human health is not a concern of agriculture, it may look like a good deal. It will remain efficient and profitable only so long as these conditions prevail—that is, so long as we eat with our eyes closed. Until these caveats are brought into the discussion, talk about returning to some kind of local or regional food system seems merely whimsical to anyone the least bit economically astute. At the moment, industrial agriculture appears perfectly capable of delivering the goods, and thus sustainable alternatives are reduced to up-scale "niche marketing" and dismissed as mere luxuries for the well-to-do. But what will happen when energy becomes more expensive, when ecological costs mount, and when we admit that what we eat may have some bearing on our health?[8]

Our ability to ship such perishable foods as milk halfway across the continent and fruits and vegetables halfway around the world is dependent first of all on cheap fossil fuel. The availability of inexpensive energy makes some very strange behavior appear economically rational. But the reckless burning of abundant coal and oil is the most environmentally problematic feature of the modern age. One of the great challenges, and great necessities, of the coming century will be to wean human society from this habit of oil dependence. As oil is depleted, it is unclear where we will find so cheap a fuel to replace it, especially for transportation. Certainly since the oil crisis of the 1970s we have improved the efficiency with which we use energy, and there are further gains to be made. Whether such improvements will allow us to continue such an inherently inefficient activity as shipping fresh produce long distances is doubtful. Whether other technological advances will unlock new fuel sources to rival the cheapness of fossil fuels is also anybody's guess. Some cornucopians assure us that our history of progress proves this to be inevitable; this historian would prefer not to bet the farm on it. In spite of the current enthusiasm for globalization, it seems unlikely that energy will remain cheap enough to maintain such extravagant global food exchange very far into the future. Global trade in food certainly will not disappear, but in order to survive, all regions will need to produce more of the basis of life for themselves.

Estimates of the rate at which new oil reserves will be discovered and then depleted vary, but it now appears that running out of oil may actually be less of a concern than finding more. This is because the more fossil fuel we burn, the more we will have to pay for the consequences of a rapidly warming climate. Some degree of global warming and increased climate instability is now a certainty. That the precise regional consequences cannot yet be accurately predicted is hardly comforting. We will quite possibly see large-scale disruptions in agriculture, including hotter, drier summers across much of western America, which would decrease grain production and reduce the surface water available for irrigating our largest produce farms. Such changes will be a direct consequence of the fossil-fuel-intensive economy of which industrial agriculture is a conspicuous and leading component—that is, the way we grow food is a culprit as well as a victim of climate change. Mainstream scientists who are hardly environmental alarmists are telling us that we need to contemplate at least a 50 percent reduction in fossil fuel consumption to forestall the worst of global warming, while our politicians struggle toward measures that might bring about a 5 or 10 percent reduction. In other words, either enduring global warming or preventing it is almost certain to bring about a wholesale change in the way we produce, trade, and consume food and other natural resources. Disrupted climate if we do nothing, drastically reduced consumption of fossil fuel if we do something—either one will force more sustainable reliance on local resources upon us. Thus we should be moving toward increased local food production—especially of those commodities such as fresh produce that cost the most to produce and distribute on a large scale—for reasons of energy conservation and climate stability alone.

But there are many other costs besides unsustainable reliance on fossil energy that are bound to catch up with the industrial food production system sooner or later. Our food is inexpensive today only because of a whole series of deferred and deflected environmental and health costs. These are not random, disconnected problems but part of a discernible pattern. In a sense, some of them have caught up with us already, but we do not properly account for them in the price of fresh eggs. They are not detected by the scanner at the supermarket checkout counter, but we pay for them all the same.

Take the high cost of maintaining the irrigation system in the Far West: it is not borne proportionally by large growers and therefore not by those who buy and consume their produce—at least, not directly. Instead, it is borne unknowingly by the same people when they pay their taxes or endure the diminished value of other things. For example, we have accepted the near complete extermination of salmon in the Columbia River watershed at least partly in exchange for a flood of Red Delicious apples, watered by the reservoirs behind the great dams. Those wild salmon, a gift of natural ecosystems that sustained Native cultures for millennia, we are replacing with less flavorful salmon fattened in fish farms. Growing salmon like that is, of course, economical only given cheap energy and cheap feed, and the industry will quickly flounder once those conditions no longer prevail—but by then the native fish may well be gone. The Red Delicious may shrivel up someday, too, but that will be no great loss because the Red Delicious is only a virtual apple: it shines but has no taste.

Or take the rapid depletion of the Ogallala aquifer that underlies the transient productivity of corn and cotton fields fed by center-pivot irrigation on the High Plains. Not only is this use of water and land unsustainable, it runs the risk of ending in desertification. Once abandoned, much of this part of the world may return not to prairie but to drifting sand, especially given a little push from warming climate. Reestablishing grass when the plains are in drought and the water has become too expensive to pump may not be a simple matter of just letting the land return to nature. Desert conditions on the High Plains are within the range of natural climate fluctuations that have occurred over the past millennium, which we seem hell-bent on replicating as quickly as we can by plowing up the grass, spilling out the water, and turning up the thermostat. With the Ogallala will go the great feedlots that dominate that part of the country.[9]

Or consider the wholesale drainage of wetlands that has created millions of acres of prime farmland throughout the Midwest over the past century. This loss of habitat has severely reduced the populations of many migratory waterfowl. It also contributes every few years to catastrophic flooding along the Mississippi and its tributaries, causing billions of dollars in damage: the hundred-year floods seem to come along about once a decade now. It may not occur to many of the people struggling heroically to clean up and rebuild after these "natural" disasters that they are paying a whopping surcharge on the inexpensive grain-fed burgers they wolfed down on their way to help sandbag the levees. Such side effects represent the high cost to society of manipulating nature to the extent necessary to maintain efficient industrial agriculture and provide cheap food. The industrial food system is really cheap not in the sense of being inexpensive, but in the sense of being glossy on the surface, shoddy underneath, and a bad bargain at bottom.

Then there is the contamination by industrial agriculture of environmental health, particularly in our waterways. Such injuries are widespread and familiar. In spite of the best engineering plans, manure lagoons lurking behind corporate livestock confinement facilities regularly breech, degrading miles of streams and filling the countryside with a stench that even farmers cannot stomach. Nothing is ever wrong with the technology, mind you: careful investigation always reveals that human error was the real culprit, a distinction lost on the fish. Pesticide- and fertilizer-laden runoff and windblown dust from thousands of square miles of farmland are leading causes of ecological disruption in our rivers and their outlets, from Chesapeake Bay, the Great Lakes, and the Gulf of Mexico to a thousand smaller estuaries. So we pay another surcharge for a cheap supply of corn, soybeans, and milk with the destruction of important fisheries and a host of what are called noneconomic damages to some of the most productive ecosystems in the world. The story is the same in Florida, where the expansion of winter vegetables and sugar cane has been paid for by the shrinking of the Everglades; the same again where the marshes at the head of San Francisco Bay serve as the cesspool for the saline residue from the ditches of the Central Valley.[10]

Agriculture is not the only cause of such biological degradation—the effluent from commercial and residential development is often as much to blame. But that is precisely the point. Suburban development and industrial agriculture often compete for the same land and water (with the suburbs inevitably winning), but these are merely two interlocking pieces of a single system of mass production and mass consumption. Each would be impossible without the other and without the cheap energy and discounted ecological costliness that sustains them both.

We also fail to properly account for the human health consequences of an industrial food system. Pollution of both surface and groundwater by pesticides and nitrates from fertilizers is a matter of increasing concern to midwestern communities and farm families, many of whom have lost their wells. The devastating impact of pesticides on the farmworkers who labor on large produce operations, both in this country and south of the border where much of our winter produce is now grown, is a humanitarian outrage. Just how much risk is posed to consumers by residues of pesticides in our food is a matter of continuing scientific debate. Concern among some researchers is now shifting from cancer to compromise of the human immune and endocrine systems. Such apprehension is regarded by true believers in technological progress as evidence of a kind of portable hysteria about chemicals among the impressionable public. I am not a particularly hysterical person myself, but it eludes me why we would allow our achingly temporary bodies to endure this kind of unnecessary experiment.[11]

Even setting the controversial issue of chemical contamination of food aside, the diet of Americans is universally admitted to be unhealthy. In other words, the way we eat makes us sick. Of course what we put in our mouths is a matter of personal choice, but such choices are themselves a consequence of inculcation from an early age. If a large part of our population persistently eats badly, this is properly regarded as a cultural pathology. We live in a culture of mass consumption of overprocessed food that is high in sugar, salt, and fat. Cheap food is plentiful in America, but good food is not easy to find. Insist on that, and you are regarded as either a health food nut or a gourmet elitist. We face a monumental task of cultural transformation. We have to deal with an industrial food system whose ill effects stretch all the way from the places where food is grown and processed to the places where people live and eat—from the earth to our own flesh.

These surely ought to be matters of concern to food consumers in New England or any urban area. The industrial food system isn't about to collapse overnight, but it is central to our deepening environmental dilemma. I am not self-assured enough as a prophet to predict that the mounting costs of scarce oil, global warming, contaminated water, and low-quality food will inevitably render this system bankrupt within another decade or two. I am highly suspicious of those who think they can prove that the end of industrial agriculture is now in sight. Such large systems are resilient and change slowly, in complex and often contradictory ways. On the other hand, I am even more suspicious of those who dismiss these costs and assure us that through biotechnology and the development of new, clean energy sources and other miracles we will be able to intensify and increase our production essentially forever, growing more and more food on ever less and less land and shipping it from a few favored regions such as the American heartland to wherever in the world it is needed. We are on course (barring some unspeakable catastrophe) for another doubling of the population of the world within the next half century. If some ten billion people are to eat well, without poisoning themselves and degrading the planet in the process, we will simply be required to grow more of our food and recycle more of our wastes closer to home. It is hard for me to imagine that in such a future there will be much surplus farmland available to be converted to some supposedly higher and better use, as has been the case in the Northeast for most of this anomalous century. It would seem prudent not to needlessly subdivide and pave farmland that is close (which now means anywhere within a hundred miles) to our major cities in our current burst of extravagant resource consumption. We would do better to limit our growth and revive vigorous local and regional agriculture.

There will certainly be nothing like complete food self-sufficiency in a town like Weston, a state like Massachusetts, or a region like New England within the foreseeable future. It is not clear to me that we would even want that. But we might do well to recreate a regional food system resembling the one we had a century ago, when we at least provided a large part of our own vegetables, fruits, milk and eggs, along with some of our meat. The population of Massachusetts has nearly doubled in the past century and the acreage of prime farmland has unfortunately diminished, so the task will not be easy. Most meat and grain will come from the Midwest (see chapter 3) and specialty products from other parts of the world. But we should be able to make market gardens flourish here again in large numbers, and long before we are forced into it by necessity.

Local farms can offer fruits and vegetables that are fresher, more healthful, and tastier than what is available in supermarkets right now, at a comparable price. The day will come when a tomato grown organically on a small farm in New England will be substantially cheaper than a tomato grown industrially in California and flown across the country. Until then, the challenge facing those of us who foresee the day when the industrial food system falters is to find ways to replant a regional food system within the current economy. Fortunately, we have not a bitter but a delicious offering on our plate. We can offer the better flavor and purity of locally grown organic food and the beauty of farms conveniently located in customers' backyards. By making direct retail connections with consumers, New England farmers can sell their produce for something close to the supermarket price of California's best effort and still pocket a decent profit. We can to some extent sell the ambiance of the farm itself: some consumers will prefer to go pick that New England tomato themselves. They may pay as much as they would in the supermarket but still feel well satisfied because they have harvested more than the produce alone. We already know this works for strawberries, and it may seep into the lower vegetable orders. When eaters turn out in droves to pull their own carrots we will know we are home and dry.

The task of pioneers in the movement for local food systems—especially of nonprofit community farms like Land's Sake—is to educate consumers by appealing to their good taste and good sense and to translate that appeal into a price that keeps ecologically responsible farmers afloat. *That* was the missing link in Land's Sake's economic equation: a direct link to our customers, right on the farm. But even having forged that, it looked as if 1985 might be our last season. We were about to lose the land.

Principles of Sustainable Suburban Market Gardening

Flowers, not food, saved Land's Sake, and possibly the Case land itself. When I started graduate school in 1983 and could no longer work full-time, the farm manager, Rob Crockett, introduced a few perennial flowers and herbs to Land's Sake. I had never had the wit to plant flowers—to me, they were a frivolity. When the arboretum finally allowed us to open a small farmstand on the property in the fall of 1984, Rob's flowers sold well. The next year, Tom Hansen of Framingham opened a larger farmstand stocked partly with produce he grew on the Case land, but primarily with stuff grown on his home farm ten miles away. He had recognized the commercial possibilities of the site and had made the arboretum an attractive offer. Reduced to a single acre and to a little stand that could sell only what Tom didn't, Land's Sake grew oddball vegetables that year: yellow tomatoes, white eggplants, apple cucumbers, and bottle gourds, tiny Jack-Be-Little pumpkins and huge Atlantic Giant squmpkins. If it was guaranteed to come up the wrong size or color, we planted it. We grew ordinary summer squash but sold only the blossoms. We grew seven or eight varieties of mint, including one patch whose seed packet was lost, which we called Experi-mint. Split open a Land's Sake watermelon and it turned out to be not red, but yellow.

But mainly, out of necessity, we grew flowers. Rob took our acre and laid it out in small beds, making the shape of a large flower at the center. He planted these beds with dozens of varieties of annual flowers that he had started in half a dozen small home greenhouses all over town. We grew flowers for drying and arranging: statice and celosia, strawflowers and salvia, globe amaranth and acroclinium, lamb's ears and Job's tears. Through a stroke of luck we planted a quarter acre of artemisia and cornered the Boston market—there was an artemisia failure out west that year, and nobody else was growing it locally. We also grew cut flowers galore: asters and zinnias, bachelor's buttons and snaps, cosmos and cleome, cheerful Mexican sunflowers and slightly sinister salpiglossis. We grew beds of towering red and white dahlias handed down from the garden of Winslow Homer, which Martha had picked up somewhere. The weather broke just right, and our garden was bounteous and intense and right by the road.

That summer was intense, too. Late in the spring, Harvard had announced they were planning to sell the Forty Acre field for development. They sent in a backhoe and dug twenty-five perc tests to see if the land would accommodate septic systems: twenty-five was the maximum number of building lots the land could support under Weston's zoning bylaw. Townspeople felt betrayed. The farm adjoined one of the busiest intersections in Weston, and we suddenly found ourselves with a spectacular garden in the eye of a storm. Land's Sake was under siege: like thousands before it, our wonderful farm was apparently destined to be disfigured by fancy houses. Weston was in the midst of a building boom, and the bulbous, inflated domiciles that were springing up dumbfounded even some of the people who bought them. Perfectly good older homes that were too small for their absurdly expensive lots were being bulldozed and carted away to mysterious landfills in Maine or were sprouting fungal additions twice their original size. Otherwise sensible people became mesmerized by the tyranny of the resale value. Settled empty-nesters added new rooms they didn't need and a hot tub they didn't want. It defied all logic except that of money. More than once, Faith and I visited apparently sane neighbors who the minute we walked in the door and gawked at their three-story marble-bottomed foyer apologized for their overblown home, behavior we found very odd indeed. They would show us their jacuzzi or their sauna and whisper, "Of course we never use it, but you have to have one these days." Now it appeared that the new Case Estates were about to rise in our humble flower garden.

We were not about to let the Case land go down quietly. With two farmstands open for business, we were now a very public place. Friends who had formerly waved to us from the road began pulling in and buying vegetables and flowers like there was no tomorrow. Tom was doing great business, too: between his stand and ours we finally had the right combination of attractions on the property, and we quietly took note of what he was selling. The place swarmed with customers. We gave them scissors and sent them out to cut bouquets, selling the flowers for a dime or a quarter a stem, a fantastic bargain. They returned the next day and cut more. The more they cut, the more the flowers grew. It seemed as though the soil itself knew this might be its last chance to bloom before being made to bear nouveaux chateaux. Gardeners know that a stressed plant will often burst into a final flowering, and those were the flowers we were selling. When the season ended, we were astonished to find that whereas the year before we had struggled to gross twenty thousand dollars on ten acres, in 1985 we earned twenty-five thousand on just one acre and made a whopping profit. Best of all, since we were leasing our acre from Harvard, we got to keep it. The hard reality of the free market had worked its miracle upon us.

But to continue growing flowers and produce, the land had to be saved by other means. The town mothers and fathers arranged with Harvard to give Weston first refusal on the property, and the drive was on to buy the land.

The price was $3.5 million for thirty-five acres—a great bargain, we were assured. This may have been something less than what the land would have fetched on the open market, but it was still the biggest land deal in Weston's history, and a lot bigger note than could ever be paid off by growing flowers. Doug Henderson and I never thought the special town meeting scheduled for the fall would vote it through—a two-thirds majority was required to issue a bond. After the property tax revolt of the late 1970s, we assumed the halcyon days of conservation land buying in Weston were over and done. It seemed like a long shot at best, or at worst a ploy by Harvard to get themselves off the public relations hook before they sold to the highest bidder.

One evening in September I was out working in the garden when I encountered a friend named Kay McCahan, picking flowers. "Somebody needs to organize a group to campaign to buy this land, now that the selectmen have cut the deal," I said. Kay was worn out at the time from the latest round in the long struggle to get the town to build a new library (which finally succeeded nearly a decade later). She stuck her finger in my chest: "*You* need to do it," she told me.

I realized that, because of what Land's Sake had accomplished, we represented a small but reckonable force in Weston. As a nonprofit we couldn't work politically as an organization, but we had friends who could. I made a few phone calls to some supporters who were influential in town affairs—Cynthia Berg, Anne Wolf, Harriet Elliston, John and Mary Lord, to name a few—and we got to work. We had all been involved in political campaigns before and knew the ropes: the flyers to be printed and distributed at the supermarket and the dump on Saturdays, the townwide mailing, the phone calls, the letters to the *Town Crier*, the full-page ad signed by friends and neighbors. We convinced even most of the real estate agencies in Weston to join the campaign to keep the Case land open; it had become central to the character of the rest of the town they were selling and reselling every day. It turned out Doug and I had been mistaken: at town meeting the bond issue to buy the Case land passed easily. The message from townspeople was clear: we don't like being put in this position by Harvard but go ahead and buy the land. To our amazement, Westonites voted to tax themselves one hundred dollars a year per average household for twenty years to save our farm.

I can't say that our farm alone saved the land. The Case Estates was a beloved place long before Land's Sake had anything to do with it: the crown jewel of Weston's open space that everyone had assumed was safe from development because it belonged to the arboretum. But our flower garden had become a big part of the land's appeal. In our five years of farming we had brought the place to life and made the Case land visible to the community in a way it hadn't been before. It had become a place where town kids spent the summer growing things, where people came to buy fresh vegetables and pick flowers. Our garden symbolized in a burst of color what that land meant to Weston. There was a great lesson for conservationists in this. Most people in Weston and even in less affluent towns can afford one hundred dollars a year and much more to save land. The question is, What will motivate them to spend their money on common land rather than on something else? Farming with kids, for one thing. I commend this discovery to land preservationists everywhere.

The next year, Land's Sake recovered tenure of all the farmland. We bore Tom Hansen no ill will—his Framingham operation continues to thrive, and we remain friends. But the Case land had become community land, we were a community farm, and we wanted our hard-won acres back. The selectmen gave the Conservation Commission jurisdiction over the parcel, although it officially remains municipal purposes land and could someday be used for a purpose other than conservation and farming. In practice, it would be difficult to dislodge what has become a town institution as long as we are thriving. Land's Sake leases the land from the town to this day, carrying out a contract to farm it to specifications drawn up by the Conservation Commission. The legal niceties are important to those of us in Land's Sake and town government who are responsible for the property, but not to anyone else. In the eyes of most of Weston, the farm on the Case land now simply *is* Land's Sake. It is the heart of what we do in Weston. It is our home ground.

Community farming and forestry as practiced by Land's Sake adhere to four main principles: the ecological, economical, educational, and esthetic. These express our attitude toward the stewardship of Weston's common land as a whole and also apply to each of our projects as it deals with a particular part of the landscape. Although these principles necessarily overlap and sometimes conflict in practice, ultimately they amount to a single, unifying purpose: building enduring bonds between people and the land by which both nature and culture flourish.

To us, *ecology* always comes first: any use we make of the land must be ecologically sustainable in the broadest and hence strictest sense of the term. First and foremost, we do not undermine or degrade the soil, water, and biotic resources from which we immediately harvest our products. But at the same time we strive to protect the overall biological diversity, health, and integrity of local ecosystems, including those that are not directly productive. Expanding the concept of sustainability still further, in the course of taking care of our own place we depend as little as possible on nonrenewable resources and industries that pollute other parts of the world. Working out the meaning of these fine-sounding precepts is a complex, imperfect business when one is productively engaged with the land and not just idly observing or preserving it.

At Land's Sake the *economic* imperative is strong but always constrained by the ecological—that is, we cannot allow ourselves to operate in the most profitable way if doing so degrades the long-term health of the land. We have to willfully distort our own market calculus. To simply proceed in the manner that costs least and pays most would inexorably transform us into what we set out to oppose. I know a back-to-the-lander in New Hampshire who started out growing organic vegetables to sell at his roadside stand, one day added a greenhouse, and eventually found himself trafficking mainly in lawn ornaments to pay the bills. "Here I am, selling *cement!*" he said to me, shaking his head. At Land's Sake, we refuse to sell cement.

On the other hand, it would do no good to go broke, either. Balancing profit and ecological principle is difficult; but Land's Sake's reason for being is to work at the crux of the problem of how to live responsibly on earth in today's economy. Our economic principle is to look for ways to work creatively within the existing market reality without succumbing to its corrosive logic. We thus accept all the public and private support we can muster and then multiply the scope of our operation severalfold by what we can earn. This means growing and selling high-quality natural products that suit our land and that have a ready local market. With a few minor exceptions, we sell only things grown in Weston.

The *educational* principle is perhaps the most important, but I put it third because it is emergent. We have found that if a farm project isn't ecologically and economically sound, then it usually isn't very educational, either. It just isn't convincing to the students to do things that require a large subsidy just to break even. We teach primarily by doing, letting the lesson take care of itself. Still, most of our projects are designed to be educational somewhere along the way. We always try to involve people in caring for the land or at least to bring them in contact with the land. This goes all the way from running the farm as a pick-your-own operation to hiring children as workers in all our projects. We find opportunities to explain what we do to our customers and our young workers, in ways that will stick. The educational component of Land's Sake may be the most important because the number of people who start thinking about how to care for land as a result of what we do is way out of proportion to the amount of land we care for. The learning is efficacious, however, mainly because the projects are large and real enough to be economically viable, and not just easily dismissible demonstrations.

Land's Sake was organized with these ecological, economical, and educational standards firmly in mind. I could rattle them off in twenty-five words or less at a cocktail party (in one of those houses with a marble foyer) or use them to frame an hour-long slide show to a garden club or conservation group in a neighboring town. Before entangling my audience in the web of details about our farm and forest enterprises, I would lay out our three guiding principles: take care of community-owned land in ecologically sustainable ways; involve people, especially kids, with this land; make it pay by selling natural products. In time I came to realize that a fourth standard had been set: beauty

In a way, the *esthetic* principle was a mere consequence of getting the other three right. But just as education could be teased out once the ecological and economic fundamentals were in place, beauty required special attention to rise above the basics. It seemed we sometimes did a thing in a certain way because it looked good or felt right. We might not always be able to put a finger on the exact ecological, economic, or educational reason.

There was an art to what we were doing, and no way around it. I will touch on these esthetic considerations at the end of each chapter because they get at how people learn to feel, to express, and to value deep bonds of affection with the places in which they live.

ECOLOGY

The ecological principle behind Land's Sake's market gardening is to practice sustainable agriculture. *Sustainable agriculture* is a slippery term; to us it means growing a wide variety of fresh food and flowers for local consumption while relying on local, renewable resources without degrading them. Our goal is to help shorten the supply lines of America's industrial food system, with its dependence on heavy doses of fossil fuel and agricultural chemicals. We aim to provide people with high-quality produce grown close to home and to keep the soil fertile through sources internal to the system and through the recycling of organic wastes. We control weeds and pests adequately without resorting to synthetic poisons. Perfectly sustainable farming may not be achievable, but we have worked steadily toward that goal over the years.

Land's Sake's market garden has grown to a twenty-five-acre farm. A few years after the Case land was saved, Weston's two community farms were merged, and today Land's Sake runs all the old Green Power projects for the town on a contract basis. Think of the combined farm as three concentric rings. The inner ring includes about two acres of intensively cultivated beds surrounding the farmstand, including the front flower garden and a side garden of herbs and salad greens. Beyond this, another ten acres or so of small fields on the Case land contain raspberries, strawberries, and successive plantings of peas, beans, summer squash, cukes, brassicas, tomatoes, root crops, eggplants, and peppers. These rotated patches are typically a quarter acre or half acre in size and serve both the farmstand and pick-your-own customers. The outer ring consists of nearly fifteen acres of the old Green Power land, two miles away in the northwest corner of town, along with crops occasionally grown on other land in Weston. Out there we grow field crops like potatoes, pumpkins, and winter squash, which supply the farmstand and are the bulk (although by no means all) of the vegetables we ship to city shelters and food pantries as part of the Green Power program.

The heart of Land's Sake's sustainable approach is to rotate crops and incorporate legumes. We have experimented with many techniques over the years and have not yet settled into a formula of farming—the real world is too complex and changeable for that, anyway. Over the years we have added and subtracted acreage, gained and lost markets. A series of farm managers have had varying ideas about soil management, relying on everything from chemical fertilizer to purchased organic fertilizer to heavy mulch and compost. But within this evolution, the constants have been multiple plantings, crop rotation, and incorporation of cover crops and green manures wherever possible.

To begin with the soil, in the early years of the Case Farm we used chemical fertilizer at minimum recommended rates of application because it was cheap and easy and we needed results. We did not have a nearby source of manure or compost or an efficient means to apply it (except for Martha, whose efficiency was infinite because she worked for nothing). But, trying to be fundamentally good farmers, from the beginning we began to put land into cover crops and green manure crops to build up the soil and wean ourselves from the chemical habit. Cover crops and green manures comprise a spectrum of crops that are grown not to be harvested for consumption, but simply to be turned under to improve the soil. Those that are grown for only part of the summer or over the winter are usually considered cover crops, whereas those that are grown through an entire growing season so that the land is out of production are usually called green manures. In either case, the cover crop serves to take up soluble nutrients and bind them into organic matter so that they are not leached away and lost but held in the topsoil. The incorporation of green manure serves as a fund of slowly released nutrients for future crops and also improves the soil's tilth and water-holding capacity; without organic matter, the light, sandy soils of Land's Sake and Green Power do not hold moisture well. If a legume is included, the cover crop fixes nitrogen from the atmosphere, and the deep, foraging roots may bring up such mineral nutrients as phosphorous and potassium from the subsoil, making these more available for succeeding crops. Finally, cover crops prevent bare soil from washing and blowing away.

The simplest cover crops are buckwheat and oats, which can be sown throughout the summer following early crops and then disked in while still young and green or just allowed to winter-kill. The classic New England winter cover crop is rye (although some prefer oats precisely because they winter-kill and so can't get away and grow too tall before being disked down the following spring). There is nothing like the sight of young rye glowing green on the fields in the low light of December, and again as soon as the ground thaws in late March or early April, to give the farmer a feeling of providence and virtue. I love to walk the fields in early winter after they have been put to bed under a blanket of rye, everything simplified and orderly. The leaves are off the trees, the grass is brown, and yet here are these bright green fields. The farm is stripped down to structure but it is still alive, and the rye reveals it. Of course, some winters resident Canada geese come along and graze the rye flat, leaving the fields bare and slippery with gooseshit; but the rye usually bounces back in the spring.

If cover crops are to really build soil fertility, they must include a legume. Whenever we can plant early enough in the fall, we mix the rye with hairy vetch, a vigorous legume that is winter-hardy in southern New England. If allowed to grow through May, vetch can provide enough nitrogen for fruiting crops like tomatoes, eggplants, and peppers. Our most reliable legumes for green manure have been sweet clover and red clover, which we often plant in rotation following strawberries. After an old patch of strawberries has finished bearing and been disked down in July, we sow clover with a nurse crop of oats. After the sheltering oats winter-kill, the clover is allowed to stand through the following growing season. Again, seeing a fallow field rich with clover gives farmers a secure feeling of good things happening underfoot while they attend to this year's crops. About one-fifth of our acreage is in a leguminous green manure crop in any given year.

Rotations protect and build the soil, but in most cases they are not enough by themselves to replace the nutrients removed by intensive fruit and vegetable cropping. Something must be added from outside the tilled land, following the ancient rule of return. Most commercial farmers simply pour on the synthetic N-P-K (nitrogen, phosphorous and potassium). As any good gardener knows, the secret of the organic alternative is to build a soil that is burgeoning with humus and microbial life (a condition far beyond the natural state of the thin, nutrient-leached soils native to New England): don't feed the crops, feed the soil. After a few years farming the Case land, we decided to give up chemical fertilizer and go entirely organic because it suited our principles better. Chemical fertilizer in moderation doesn't harm the soil, but it is nonrenewable and too often substitutes for good soil and waste management. A simple alternative is to purchase bags of organic fertilizer. These mixes are mostly recycled by-products from the industrial food system—commendable and handy for some but hardly a universal solution either. What we really need is more compost and manure from local sources, something Land's Sake has never had enough of, mainly because we have never had a sufficiently large livestock herd or a barn in which to collect the manure, in the manner of market gardens a century ago.

The suburbs can supply market gardens with plenty of compost. All summer long the landscapers mow lawns and haul away the clippings, and in the fall they vacuum the leaves that fall on those lawns into their high-sided trucks. For years these residues piled up into a mountain of black mould in an old gravel pit on the Weston College land. We spread some of that pile on the nearby Green Power fields, but we grew wary because we found all kinds of garbage in it. In principle, using municipal organic wastes on local farmland makes great sense. Ideally, this would include well-composted food waste and human waste. In practice, our society often fouls its wastes by adding so many toxic substances along the way that organic farmers are naturally leery of what comes out of urban composting operations and treatment plants. In the future, when this culture becomes concerned enough about not poisoning itself and not squandering its soil fertility, perhaps we will think more carefully about the flow of nutrients at the community level. Weston now has an orderly composting operation for yard waste. In 1997, Land's Sake resumed spreading leaf mould on its gardens, with good results.

Another source of local organic matter for some of the more intensively managed parts of our farm is mulch. We mostly use spoiled hay from nearby horse farms, some of which inevitably gets rained on or cut too late every year or is left over in the spring so that we can buy it cheap. We use hay to mulch the strawberries through the winter to keep them from heaving with the frost, and later to keep the fruit clean in June. We have also used hay and straw to mulch the front flowerbeds, and some of the other crops from time to time. Delivering nutrients is not the primary purpose of mulch, but it does have that added benefit over time. Its primary purpose is to keep the soil moist and, if applied heavily enough, to control weeds. Deep mulch works well, but applying it is very laborious. Mulching, or sheet-composting, by hand is a good way to run an intensive market garden of only a few acres, and we have usually run part of our farm that way. The inner ring near the farmstand gets most of the compost and mulch and responds with the best crops. Returning organic matter to the land in bulk requires equipment and a great deal of time and effort. Our diverse program pulls us into other projects during the farm off-season, so we have been slow to take up mulch and compost as devotedly as they deserve. We still buy a good deal of our fertilizer in bags or jugs, albeit ones marked organic. Land's Sake has been free of chemical fertilizer since the late 1980s, long enough to be certified as an organic farm by the New England Organic Farming Association (NOFA).

Regarding pesticides, Land's Sake has long held to a strict organic standard. Our pest problems have been tolerable. Through my farming years I have lost plenty of sleep over politics and personnel, some over frost and drought, but none that I can recall over insects. By making a wide range of small plantings and rotating them constantly, we manage to spread our bets and avoid major pest outbreaks. We expect to write off a few plantings every year for one reason or another, depending on growing conditions. Crops that get off to a slow start are always susceptible to pests, and our practice is to just disk them in and have another, better planting coming along. We have had misadventures with Mexican bean beetles, corn earworms, striped cucumber beetles, flea beetles, leafhoppers, cutworms, pheasants, crows, mice, voles, groundhogs, raccoons, and deer. I am expecting an infestation of black bears to reach Weston any day now. I imagine there are ingenious organic ways to thwart most of these pests, and we have tried many of them, sometimes with success, sometimes not. Some biological insecticides work very well. *Bacillus thuriengensis* has proven a godsend for controlling cabbageworms and Colorado potato beetle. We formerly used rotenone to knock back cucumber beetles but became concerned about the reputed danger of that botanical toxin to the human reproductive system and swore it off. For the most part, we rely on sufficient scale combined with high diversity, take our lumps, grin and bear it. I doubt our pest losses are much higher than those of farmers who spray for everything in sight, thus kicking off entirely new infestations of some obscure mite previously unknown to agronomic pathology. I guarantee our costs are much lower. On a commercial truck farm, a high percentage of many crops doesn't make the grade and never gets sold in any case. Pests don't bother us much.

Everything in the Land's Sake ecological approach to market gardening finally rests on crop diversity. Of course we rely more on some crops than others, but even among the most reliable crops we never know which will have a big year and which will disappoint us. So much depends on the weather, the timing of rain and frost. Our philosophy is to get crops in the ground early, keep putting them in as late as we can, and then respond in a timely manner as the season develops. When the weather is wet, we go out in the mud and transplant until it is too dark to see (we can get away with this in our sandy soils). When it is dry, we cultivate early and often. When there are crops to pick, we pick them in their prime. Our approach may seem a bit slipshod and haphazard to those who make a huge investment in a single crop, but we always seem to have plenty of something to sell. We seldom have a serious failure of the big money crops such as strawberries, tomatoes, and pumpkins.

It doesn't do much good to build soil fertility and get healthy plantings if you do not control weeds. I have little patience with theories of minimal weeding. We learned the hard way that it is far better to cultivate before the untrained eye can even see the weeds than to wait too long. As soon as those spindly stalks with two tiny leaves make a thin sheen across the soil surface, panic. Take a gang of kids with hoes and wipe them out. If you wait, a rainy spell may come along and put you deep in the wilderness. Amaranth, lambsquarters, and ragweed can go from a quarter inch to four inches tall overnight, and then you won't hoe so fast. You may cultivate an extra time or two in the course of the season by striking early, but you go so much faster through the loose soil that it takes less time in the end. Unless you are operating on a

John Santos cultivating everything cultivable with a rototiller, 1986. (Kippy Goldfarb/Carolle Photography)

small scale where you can work very intensively with mulching, companion planting, and picky hand-weeding, there is no substitute for clean cultivation.

As for our old nemesis, witchgrass, we finally triumphed with pumpkins. "Convince me that you have a seed there," wrote Henry Thoreau, "and I am prepared to expect wonders." As his metaphor for the growth of ideas and morals he used a squash seed, but it might as well have been a pumpkin. To me, pumpkins are a truly wonderful crop. We learned to use pumpkins to clean our grassiest fields, sometimes with the old sod not even completely rotted down yet by planting time. Like any farmer, I love an activity that accomplishes two things at once, and growing pumpkins to kill witchgrass definitely qualifies. We plant pumpkins in rows twelve feet apart so we can run the disk between them. A few pounds of seeds are enough to cover an acre or two. Believing that in a few months an entire field will be a sea of pumpkins and nothing else always seems an absurdly optimistic fantasy, but it never fails.

Once the sturdy little pumpkin seedlings show themselves in the spring tide of witchgrass, it is time to disk regularly between the rows to beat on the grass and to hoe in close around the plants. The best way to eradicate perennial weeds is to exhaust them by constant harassment—force those rhizomatous roots to keep pushing up new shoots until they grow weary and die. By midsummer the pumpkins have grown big leaves and tendrils, the grass is looking tired and thin, and the whole gamble suddenly seems inspired. Then the pumpkins run, their roots well established in the dry soil, and any live witchgrass is smothered, desiccated, and destroyed. Big yellow blossoms appear by August, one day we notice the mottled green fruit beneath the waves of leaves, and by September the miracle is complete. The pumpkins turn deeply ribbed orange and litter the ground in profusion as the vines turn yellow and die back. They are a trouble-free crop and cost next to nothing to grow. One Saturday morning we recruit a gang of friends and children and line the pumpkins up in rows, back the truck across the field tossing pumpkins to catchers on board, and unload them in a big orange pile by the stand, under the Norway maple. In October, people pay us good money for them—they are among our most profitable crops. By Halloween the pumpkins are sold, the field has been disked smooth, the witchgrass is gone, and the first purplish blades of winter rye are coming up. Pumpkins complete the circle of crop rotation from soil-building green manures back into clean cultivation.

Gradually, sometimes painfully, we learned the tricks that seem to work best for our soils and climate. Such slow accumulation of minute particles of experience makes farming rewarding. Knowledge builds up in the mind in much the same way that organic matter builds up in the soil, as the residues of one year after another are turned under. Thus every year we cultivate the ground not only in the present, but within the memory of what happened with this crop or in that field in past years; things to watch out for, things to take advantage of. Deciding that the season has been rainy enough to try an early planting of beans on a gravelly knoll where they would shrivel in a dry summer but will warm up faster in this wet one. Deciding against a late planting of beans in a small pocket nearby and putting in a few rows of carrots instead, because we vividly recall the flourishing eggplants that hung there brown after an early September frost. Those who find farming monotonous must have their minds off the premises or have oversimplified their operation in pursuit of the bottom line only to discover that the bottom line is drudgery. Then they declare that those of us who say we farm for enjoyment are hopelessly naïve or mere hobbyists—farming is a business, they insist, and the rest is romantic hogwash. But the familiar repetition, always with small variations, is precisely what makes farming so engaging. A farmer is like a musician playing the same symphony with a new conductor every year. In the end, this deepening familiarity with a place makes farming sustainable.

ECONOMY

But farming is a business, too. By the late 1980s, after a number of false starts, the Case Farm began to show a consistent profit. Since then, it has usually run in the black, helping to support other parts of the Land's Sake program. We are not unusually talented growers—we are still neophytes and likely to remain so because as a nonprofit organization we have a dominant neophytic gene. Our methods are inefficient and our yields are often mediocre, and still we generally come out ahead. This is important because it shows that suburban market gardening really can pay, indeed can hardly help paying, once the exorbitant cost of suburban land has been removed from the balance sheet. If Land's Sake can do it, anybody with a little ingenuity and a lot of gumption can do it.

Our modest economic success derives from our diverse, organic approach and from the way we deliberately attract customers to the land. A clean chemical grower might be able to produce more tonnage on the same land, but I'm not sure they could sell more at a better price, with so little marketing effort. A really good, intensive organic grower would earn twice as much as we do. We farm with kids, and it shows. Still, the spirit of what we are doing pervades the place and is definitely reflected in steady sales. We seem to have arrived at a good economic formula for community farming.

Our farm is a public place. It is a joy to visit, if not to buy vegetables then to walk among the small fields and groves of specimen trees. The Conservation Commission and Land's Sake welcome walkers. Our farm is located within a few minutes' drive of several hundred thousand people, next to a busy intersection heavily traveled by commuters and suburbanites out shopping and ferrying the kids around. Therefore, we rely mainly on the farmstand and on pick-your-own customers to move our produce on-site. We rarely visit farmer's markets or sell to restaurants, as many successful growers have done. Neither have we tried the community-supported agriculture approach of supplying weekly bags of produce to prepaid subscribers. These methods of bringing farmers and consumers together work well for some, but they have not been necessary for us. We do give Land's Sake members the opportunity to purchase season "picking passes" (which they seldom utilize to full value), and this provides a little spring flush of cash. Land's Sake is a community farm by definition, so we entice our customers to come out to the farm. This approach might be less attractive to private growers whose farms are also their homes.

Flowers remain our most valuable crop. An acre of flowers draws people off the road and floats them away into a kind of horticultural dreamworld. They sometimes get lost out there, awash in waves of colors and scents, surrounded by the hum of bees. Many of our customers come several times a week for the pleasure of cutting fresh flowers for their homes. Flowers need to be cut to keep producing, so we sell most of them at a low price. As a result, customers pick large bouquets and come back often. We have a large selection of annual flowers and have gradually added perennials, which give earlier blooms but do not produce as prolifically. The perennials also save us from the chore of replanting the entire acre in small beds every year. Besides handing pick-your-own customers the scissors and a cutaway plastic gallon jug with a little water and letting them go at it, we also gather bouquets to sell at the stand ready-made for a higher price. The flowers are profitable, and they set the tone for the whole farm.

In the late 1980s, we had a flourishing dried flower operation. We hung the flowers in colorful bunches to dry in a small barn nearby and supplied several local flower shops. Our mainstay was artemisia, which is used as the foundation for many wreaths and arrangements. Artemisia is easy to raise but stressful to harvest because you have to cut it about a day before the silver buds burst and become worthless. Artemisia is a simple and lucrative

Woman picking dahlias at the Case Farm, 1985. (Jerry Howard/Positive Images)

crop but seems to induce erratic behavior among those who handle it—maybe it's that pungent aromatic dust that comes from the leaves. It always left me lightheaded and probably contributed to the manic atmosphere of the annual artemisia harvest.

After flowers, our most important cash crops are strawberries and raspberries. We usually have about two acres of strawberries in production. Strawberries get the farm year off to a good rolling start and earn us some money after months of draining spring expenditures. Most of our berry business is pick-your-own, although we try to keep the farmstand stocked with quart baskets when the season is at its height. Getting people out to pick fits our philosophy, and it is good business as well. Many of our customers are families with young children, and we want them to enjoy the land and keep coming back. The berries have not been sprayed, so the parents don't have to worry if their kids eat a few in the field. We don't bother much over losses from sampling and trampling or try to keep customers within roped-off areas like some farms do. We charge a high price for our organic berries and let the folks go at them. People have a good time and are happy to pay for it. They often insist on paying for the extra ones they ate while picking. We also do well with our rhubarb patch during strawberry season.

Land's Sake grows some blueberries and summer raspberries, but fall raspberries do better on our droughty soils and so provide our second major crop of fruit. Fall raspberries, introduced by plant breeders a couple of decades ago, revived the fortunes of raspberry growers in our part of the world. In 1920, there were 481 acres of raspberries in Massachusetts. By 1969 the number had fallen to only 22 reported acres—but by 1992 raspberries had rebounded to 140 acres! The miniature explosion in raspberries is fueled mainly by fall varieties, which bear on the first-year canes. Managing them consists of merely mowing them down after they finish bearing, and they are much less troubled by disease and cold than their summer cousins that overwinter above ground. People love going out to get raspberries as the weather turns a little cool in September, and they quickly pick the bushes clean. In an extraordinary year we have sold raspberries into November. One year I picked a few at Thanksgiving.

A remarkable transformation comes over the farm every year around Labor Day. Families are back in town, the kids are back in school, and the atavistic urge to gather in the harvest seems to swell in the suburban breast. Business picks up by late August, and then on fall weekends it balloons. The Case Farm is thronged with customers. Having heavy fall crops is crucial to a successful farm year. If we get warm weather and hold our tomatoes into October, we're sitting pretty. A few people like to pick their own vegetables, but most of this business is done from the stand. We keep the shelves and benches stocked with produce, crank up the cider press on weekends, and rely on the fall flowers and raspberries to bring people into the farm. Heaps of late tomatoes, several varieties of winter squashes, and the giant pile of pumpkins make the place an overflowing, colorful cornucopia. It is an exhilarating time of year because of its bounty and because risk is in the air. After six months of driving labor to stay on top of the season, the financial success or failure of the year rests on a few weeks, on when frost strikes, and whether the weekends are foul or fair.

Part of running a successful suburban farm is admittedly offering rustic entertainment—though there is always danger of going too far. Like many farms, Land's Sake puts on a Strawberry Festival and a Pumpkin Festival in the course of the season to bring out the crowds, featuring meals of farm food, hayrides, a few lambs in a pen, craft demonstrations, storytellers, and music. Every few years the *Boston Globe* makes one of these events their "Pick of the Week," and thousands of people descend on the stand and clean us out. When a celebration pulls in more cash than a week of normal sales, I am left feeling exhilarated but uneasy. Why is it that people want their food to be dirt cheap but will gladly drop a few bucks to hang out at a farm happening? The festivals help pay the bills, yet I remain wary of the day when we find that growing the food is incidental to running a petting zoo and fun farm.

One of these events is really special. Every year at the end of September we hold a Harvest for Hunger at the Green Power fields. Throughout the season we send part of our fresh organic produce to Boston's homeless shelters and food pantries. At the Harvest for Hunger dozens of volunteers from Weston and beyond turn out to help get in the bulk of the potato and squash harvest, and thousands of pounds of Weston vegetables ride the Boston Food Bank truck into the city. We run a "pickathon" that works like a walkathon: kids and adults solicit sponsors to kick in a few cents for every pound of squash or spuds they put in their burlap bags and carry to the edge of the field. This helps us cover the cost of growing the food. The Town of Weston pays Land's Sake wholesale prices to send at least twenty thousand pounds of produce annually to inner-city (and a few suburban) food programs. We often ship considerably more than we are paid for, so we make our contribution, too.

Sending food to shelters is an effective way for Land's Sake to move surplus to an assured wholesale market, and amounts to about one-fifth of our annual farm sales. Having such connections to institutional buyers and low-income markets is an important part of any good local food system. Charitable organizations that provide food to the needy can support local growers at the same time. Growing food for both the high-priced organic retail market and the food pantries makes a nice balance for Land's Sake.

EDUCATION

Young people are a big part of everything we do at the farm. Now children are not the world's most efficient workers. On the other hand, they have always worked on family farms and in agricultural communities, which is healthy for them and handy for the farmers. Child labor was the basic means of cultural and economic continuity in our agrarian past. Within reasonable limits, it might have the same utility for us today. We can, if we prefer, go on lengthening the school year so that young people will be properly trained to switch jobs and migrate every few years of their lives—and watch our communities continue to fall apart. Better to get them out of the classroom and back on the land, I say. Better to shorten the school year, not lengthen it. I say this seriously, knowing full well that almost nobody will take it seriously. I believe that it really does take a village to raise a child, but the global village is fictitious and cannot teach people to care for the places where they live. What we need to raise children are real villages.

At Land's Sake we like working with kids, and I do mean working. We do mechanical work with tractors and rototillers and, for a few years, draft horses; but we also do more handwork than an average truck farm in order to involve the kids. Young people transplant, hoe and hand-weed crops, stake up tomatoes and lay down mulch, and pick, wash, and sell produce. In the early years of Land's Sake the crews were made up of a few motivated high school and college kids. At the beginning of each season we quickly winnowed the new crop of farmworkers down to those who really wanted to be there. We tried to pay our workers a wage for *growing* food equal to what they could make bagging groceries or flipping burgers, which wasn't always easy in the boom years of the mideighties. We selected young people with an appetite for farmwork and ran a streamlined operation.

After Land's Sake absorbed the Green Power program, we began to work with middle-school students. Summer activity for kids this age was an established part of Green Power, and now we had to accommodate that role on a farm with a far stronger commercial imperative. The Conservation Commission subsidized the younger workers, and in return we built education more deliberately into the program, adding a weekly field trip: from swimming at Walden Pond in Concord to helping prepare the food they helped grow and serving lunch at a shelter for homeless women in Boston. The young Green Power workers are at the farm from eight in the morning until noon. Older, more experienced hands serve as crew leaders and in the afternoon swing into tasks unsuited for little kids. Those who like the work can rise through the ranks as they grow up. I can see why private commercial growers are impatient with students and prefer hardworking migrants, but on a community farm kids are an appropriate labor force. The Green Power program is popular, as many as one hundred children participating each summer.

Hidden economic benefits of farming with children in this benevolent way go far beyond access to "cheap labor," which by itself would be a losing proposition anyhow. The parents of the children become loyal Land's Sake supporters, giving us substantial membership contributions every year. They also become steady customers, as do many other people in Weston who recognize what we are doing for the community and patronize our stand for that reason. And the attraction extends beyond Weston: seeing all those children charge out to the fields hoes in hand makes our farm a place people love to visit. The same holds true for the family appeal of our pick-your-own approach to harvesting. Families come not just to see farming going on, but to take part in getting their food from the soil. Many families in Weston provide Land's Sake with workers, customers, and underwriters all at once—and as Weston citizens they own the land, too. People are hungry to make these connections, especially when it comes to their children. Farms that can respond in simple, uncontrived ways will do good business.

The educational program is not overplanned. Parents like their children to "get their hands in the soil" and "learn where their food comes from," and we get a lot of support for that reason. Parents love it when the children come home all covered with dirt and so proud of themselves. That's real education. We don't prioritize the core farm experiences in which every child must participate to internalize a holistic appreciation of the food-growing process. If we did, the place would wither on the vine. We grow the crops that do best in our soils and climate and for which we have developed a market. We employ the kids at tasks for which their skills are suitable, and because of that what they do is meaningful. People they know stop at the farm and buy the stuff they grew, take it home, and eat it—now that's positive reinforcement.

Of course, not all Weston children like farming—some are there only because their parents think it's a good idea. At times it comes home to us that in fact we are running a cut-rate suburban day-care service. But for the kids who do like it, we find tasks that demand responsibility and engagement. We work alongside them and talk about what we're doing and why, without being overbearing about it. When it takes, we generally find we are echoing a message they're getting at home. But talking about good environmental behavior is one thing, having the opportunity to get out on a farm and learn to take care of land in your own suburban town is quite another. A small but steady stream of Weston young people, maybe one or two every year or so, come through our program and actually do go on to become farmers, horticulturalists, or natural resource managers. Some come back to work full-time for Land's Sake during their twenties, then take up farming and forestry on their own somewhere else. I like to tell people that they are our most important crop, and in my heart, I believe it.

ESTHETICS

Land's Sake is a farm in a fishbowl. We occupy conservation land where people are welcome to come for a walk and expect the place to look beautiful and well-kept, as they should. But more than that, we sit by a busy intersection, and Newton Street follows high ground along the south side of the farm, giving motorists sweeping vistas into almost every corner of the thirty-five acres within which our fields are set. To most of our fellow Westonites, we are viewscape. Everything we do is in plain public sight. "People judge us according to how we look," we tell the kids. "Most people in this town don't know anything about farming, and unfortunately a lot of them will never even set foot in here. What they can see from the road is all they will ever know about us." Keeping up appearances is a never-ending battle when you have a farm full of young people.

The way the farm looks is enormously important. Beauty is one of the best things suburban community farming can offer. Most people who visit Land's Sake overlook the occasional superficial clutter of a working farm and see only the striking beauty of the place. We are blessed to be farming not in a big, square field, but in a lightly rolling landscape graced by small woodlots, clusters of flowering trees, and grand specimen beeches, maples, and spruces more than a century old. We have worked within the lovely pattern we were granted and considered it a privilege to be so pleasantly inconvenienced. The colors and lines of our flowerbeds and crop rows add new touches to that intimate landscape, changing with the seasons. Our care makes a place that people love to pass by, and some do stop and walk around. This suggests to me that growing fruits, flowers, and vegetables at this scale—larger than a backyard garden but smaller than a mechanized truck farm—has the potential to create gardenscapes of common beauty throughout the suburbs that surround our cities.

I know planners have been advocating garden suburbs for a long time, but I haven't encountered many effective mechanisms for getting them to happen. Market gardening could provide the means, if done with the right sensibility. I have seen other small market gardens in settings that are just as beautiful as ours, if not more so. Given the healthy urban market for fresh fruits, herbs, and flowers, there are endlessly varied ways to compose landscapes of trees, shrubs, and gardens that are intensely beautiful as well as profitable. Many of these market gardens will be privately run, but the esthetic considerations are especially apt for public land. Once suburban land has been protected from development, it needn't cost another dime to maintain it by market gardening because the harvest pays for the upkeep. The machinery needed to farm in this way is no more disruptive or dangerous than that used to maintain a city park. If the farm is run organically, it is probably a safer place to walk than the average park, which is periodically soaked with poisons. An intensive, productive use of land in which the economic and the esthetic are deliberately fused is no fantasy: it is being practiced now, and it works. It expands our notion about having an engaged, attractive relation with the land beyond the limits of the home garden and yard.

Vision is not the most important sense to be gratified by local market gardens. It is the flavor and healthfulness of the food grown in this way that most recommends it. Good eating, that's the point. The purpose of growing produce close to home is to eat the best possible food. Every gardener knows that green vegetables eaten immediately after they have been picked are nutritious and flavorful. Beyond that, produce grown and marketed locally can be harvested at the peak of ripeness instead of underripe, so that it will keep, or overripe, so that more of the crop is mature when the big picking machine goes through. You can enjoy young beans and zucchini so fresh that, quickly steamed, they squeak against your teeth when bitten. You can savor unbruised and unspoiled leaf lettuce and other tasty greens that haven't been rotting under a supermarket mister, a few ripe cherry tomatoes, and maybe a handful of bright nasturtium flowers. Beyond freshness, local produce lets you enjoy varieties that are bred for their flavor, not for their ability to remain stiff during grueling transcontinental voyages. What could be more elementary? But the truth is that most people in this supposedly affluent country rarely eat meals prepared even partly with such superb food. It is odd that we routinely skip over something so basic and so simple to satisfy in our pursuit of the good life. A hundred acres of garden land in every town would put this right.

Growing and eating good food satisfies the unity of the senses. The beauty of the landscape gets into the flavor of locally grown food. That is, if you know the land where your food is grown and can recall it (or even see it) as you eat, and it is beautiful, the food tastes better. If you have had a hand in growing it, the food tastes better yet. Gardeners everywhere know this is true. You may say that the food doesn't really taste any better, but of course it does. Taste is not merely chemical. Flavor may start in the mouth, but the full experience of food is finally put together and appreciated in the mind. The same mouthful of food tastes different depending on what else we have been eating with it, how hungry we are, how tired we are, what our mood is at dinner, whom we are eating with. Knowing where the food comes from should be part of the normal appreciation of an everyday meal. Eating in this way is a celebration of the land and of the care of the land. What looks most beautiful should taste most delicious, connecting the place to the palate, joining flavor to reflection.

I don't want to make a fetish of this. Take too many steps down this garden path and you wind up in aromatherapy. I am not a zealot who believes that *only* by eating strictly locally grown food can we be truly healthy and at one with our native region, or that *only* by eating in season can we come to be biologically in tune with the circle of time. As I have said, I am not claiming that we could or should grow all of our own food in Massachusetts, let alone Weston. There are plenty of delicacies from other parts of the world that I want to continue to enjoy, plenty of staples that it makes sense to grow in bulk in other regions. I am willing to account for the full ecological costs of producing such foods sustainably and shipping them to me and to pay for them if I find they are worth the price. I am not the ecological economist to determine exactly how much of which foods should be grown in New England. But I am convinced that Americans could grow far more of their food locally and benefit greatly. The food grown in this way will generally be tastier and healthier, the agriculture that produces it will be more sustainable, and in the process we will gain both a practical and a profound connection to the land we inhabit.

Market gardening is a good place to begin reviving some semblance of a regional food system. The superiority of fresh local fruits and vegetables is compelling even in today's cheap energy economy, and the demand for organic food steadily grows. Such produce can be directly marketed quite competitively by small, intensive operations, even as small as five acres or less. Many pieces of land in the interstices of the spreading suburbs, too small to be farmed efficiently by modern, conventional means, can grow organic produce at a profit. Much of the best tillage land near the city has unfortunately already succumbed to suburban sprawl, but because plenty of organic matter is available and markets are close at hand, the native quality of the soil is not all that important—almost any scrap of land will do. Gardening is an intensely modified use of land that takes place near habitations where organic matter can be profitably concentrated—there are people making a living with raised beds on rocky mountainsides overlooking ski resorts and in the rubble of inner-city vacant lots. There is no reason why the suburbs of every city in the country should not be colonized by small market gardens, creating a new gardenscape—if the remaining open land can be protected, even in very small pieces.

Reviving this part of a local food system would require a trivial amount of land. In 1910, market gardening in Massachusetts peaked at about 40,000 acres of small fruits and vegetables, and then began a steady decline—except for a suggestive jump back up to 45,000 acres during World War II. Today we have fewer than 20,000 acres, but meanwhile our population has nearly doubled in size. Could 40,000 or even 80,000 acres of market garden land still be found in Massachusetts? Easily. The census reports there are currently 190,000 acres of active cropland across the state, and this does not count the thousands of smaller pieces of idle land within the broad suburban fringe. Only a few hundred acres of market gardens on average in each town would be required for market gardening to regain and even surpass its former glory. Weston is a suburb of medium density, and I know where to find well over 100 acres of tillable land in our town. The suburbs and surrounding rural towns that are just beginning to feel development pressure could again provide fresh vegetables in season for a good part of the population of metropolitan Boston.

If the land can be found in a crowded state like Massachusetts, where good tillage land was never plentiful, it can easily be found elsewhere. Many cities—Washington, Philadelphia, Chicago, and Kansas City to name a few—are sprawling onto much less cramped hinterlands composed of prime farmland. The land is there, in theory. It only needs to be protected from development by more rational rural resettlement patterns than dispersed tract housing or, worse yet, five-acre ranchettes. Vegetable production (not counting small fruits) in Massachusetts bottomed out at 15,300 acres in 1982 and had risen to 16,600 acres by 1992, an encouraging trend. If this turnabout is to continue, we must get much more serious about protecting remaining suburban farmland.

The best tool for protecting farmland may be the conservation easement. By purchasing development rights, the community makes it possible for private farmers to stay on their land and farm. I have not been suggesting that most of the produce grown in Massachusetts should be grown on community farms. The expanded market gardens of the future will be run mostly by private farmers, just as they are now. Much of the *land* will need to be protected by conservation easements, or we will lose it. By acquiring easements, a community assumes part of the burden of keeping farmland from being developed. This strikes me as an appropriate public investment by citizens who would like to live among small farms. I will have more to say about protecting land in the final chapter of this book.

I know from experience that it takes many years of farming to learn what works best on a given piece of land. Centuries of experience are even better, which is why small family farms are ideally suited to the task. Unfortunately, those generational chains are almost all broken now and need to be forged anew. I would like nothing better than to see some of the hustling young landscapers who now flourish in the suburbs turn themselves into market gardeners. They have the necessary horticultural skills and entrepreneurial drive. Indeed, many of them are descended from market gardeners of only a generation ago. It is a shame to see such talented people mowing lawns and planting gaudy rhododendrons in floating islands of bark mulch around garish suburban palaces, when they could be growing splendid food. I know landscapers who would a thousand times rather be running a small farm if they thought it would pay and if the land were available at a reasonable cost. It will be a great day for this country when landscapers return to growing food and flowers in intensive gardens and greenhouses, and the residents of the suburbs devote as much attention to surrounding themselves with productive gardens as they now lavish on merely ostentatious grass and shrubbery. That day will come if we protect good land in our communities and pay good money for food grown on it.

Even such a flowering of small private market gardens would leave a special place for at least one community farm in each town. Nonprofit, educational enterprises like Land's Sake may never be as efficient at growing food as a well-run family farm, but they have an equally important role to play. Community farms serve as catalysts for a broad revival in market gardening, supporting the private sector. Community farms occupy public land and so are perfectly suited to bringing customers onto farms and exposing them to the multiple benefits of local food production. They educate children and their parents about growing good food and caring for land. In this way they produce not only newly aware consumers, but also new farmers, which is what we need most of all. Community farms like Land's Sake teach people about market gardening, build demand for fresh produce and thus increase the market, and generate broader support for protecting farmland and for farming in general. As long as they do not take advantage of their public subsidies to undersell private farmers (which they would be foolish to do), community farms will not compete harmfully with private growers. They help create a more supportive environment for suburban farming. Land's Sake graduates have started their own small farms in the area, and we are proud of them. I would love to see dozens of new private market gardens spring up in Weston and surrounding towns, for which Land's Sake would help prepare the workers, farmers, and customers.

Market gardening is preeminently a suburban form of farming. It requires large nearby markets and can accommodate large concentrations of organic wastes—the recycling of nutrients that enter the region partly in the form of other foodstuffs from the grain belt at the nation's midsection. Gardening creates an intimate landscape in the immediate vicinity of human dwellings, surrounding the village, if you will. For these reasons, the denser inner suburbs will naturally have community farms that focus on gardening rather than on livestock or forestry. Community farms in more rural areas may need to strike a different balance. Land's Sake is somewhere in the middle: market gardening provides about half of our income, but it occupies only about 1 percent of the two thousand acres of conservation land with which we are concerned. Forest is the mainstay of our common land in Weston, and I am sure that the protection of forestland will be the main function of the reclaimed commons in many rural towns. I will come to the forest later in this book.

But what about the land that lies in between pumpkins and pines, the hayfields and pastures that once gave New England its character? Community farms may have a role in reclaiming this part of the landscape as well. Livestock should play a part in every community farm program, and that is what I will discuss next: grazers and grass.

Faith Rand with two ram lambs and their mother, 1986. (Kippy Goldfarb/Carolle Photography)

Chapter 3

Livestock and Grass

At first blush, keeping sheep in the suburbs seems about as foolishly atavistic as it gets. In today's world, lamb and wool are produced by large flocks on the fringes of civilization—in the Australian Outback, Outer Mongolia, Montana, or anyplace where rangeland is relatively cheap. The small farm flock is an anachronism anywhere in the northeastern United States, let alone foraging over some of the most expensive grass in the world: suburban backyards ten miles west of Boston. If all the sheep in Massachusetts were subtracted from a single Wyoming flock, they would hardly be missed.

There is a story about a conference on sheep-guarding dogs held at Hampshire College in Massachusetts. Two western ranchers who each owned tens of thousands of sheep were chatting with a woman from Vermont, who told them she had one of the largest flocks in New England.

"About how many head do you run, ma'am?" one of the ranchers asked politely.

"I have over a hundred breeding ewes," she beamed.

The ranchers glanced at one another. "Oh?" responded the second one. "And what are their names?"

In the late 1980s the Russians and the Chinese stopped buying wool and mutton, and the value of sheep on the world market dropped so low that ranchers in Australia began herding their flocks into pits and shooting them by the millions. After reading about this, I looked at the little flock of thirty-five ewes upon which Faith and I lavished such care and wondered who was crazier, the Aussies or us. In a world with millions of surplus sheep, why were we bothering with a few dozen? Then again, how could they do that? I could grasp the economic logic of such useless slaughter, but beyond economics, logic failed me. It occurred to me that it was obviously better for sheep to be kept by economically irrational people, and I immediately felt much better myself.

Land's Sake got into the sheep business because Faith wanted to raise wool and because a barn and pasture presented themselves to us. Through the 1980s we lived in Weston's remote northwest quadrant in a small house on the edge of Jericho Forest. Down the lane was the remnant of a colonial farm—a handsome eighteenth-century farmhouse, a picturesque nineteenth-century barn, and four or five acres of vacant pasture. The place belonged to a pediatrician, Ralph Earle, and his wife, Shirley. The Earles' children and their horses had recently left home, leaving behind not only an empty nest but an empty barn as well. Faith and I proposed a neighborhood sheep partnership.

Over the next few years we roped a dozen small pasture owners in Weston into this scheme. Each had a few acres of rundown field out back they had no desire to maintain as lawn but did want to keep open. Their part of the bargain was to donate money to Land's Sake to purchase ewes and the portable electric fencing to confine them. In return for this they were privileged to gaze upon grazing sheep over their evening cocktails. Faith and I set up the fences, moved the flock from pasture to pasture, and came by to water them daily. We did not charge Land's Sake for our time—Faith got the wool for her trouble. In the winter we brought the flock home to the Earles' barn. We lambed in the spring, raised the lambs on grass and mother's milk, and had most of them slaughtered in the fall. Sales of freezer lamb and tanned sheepskins paid for hay, a little grain, and veterinary supplies. Those were the main lines of our sheep enterprise.

We were not alone. During the 1980s, there was a sheep boomlet under way in New England, a faint echo of the great sheep explosion of the early nineteenth century. According to the census, the number of sheep and lambs in Massachusetts more than doubled from an all-time low of 6,515 in 1978 to 14,761 in 1987. The Middlesex County flock nearly quadrupled from 664 to 2,302 during the same period. By comparison there had been at least 190,000 sheep in Massachusetts toward the end of the real sheep boom in 1850, but a mere 1,844 of those were in Middlesex County, which remained cow country.[1] By the late 1980s there were more sheep in Middlesex County than there had been in more than a century. I know, I know, and their names were Daffodil, Ambrosia, Muffin, Fanny, Quiet, Baby . . . but at least things seemed to be moving in the right direction.

What brought about the revival? For one thing, improved electric fence, which cost less than wire-mesh fencing and still protected the stock from the depravations of dogs, not to mention the fast-emerging eastern coyote. For another, enthusiasm about Finn sheep, which have a propensity for multiple births. A little Finn blood in the flock increases the number of twins, which means a higher return for each breeding ewe that is carried through the year. These factors, along with growing interest in rotational grazing, which the new movable electric fencing facilitated, made it seem plausible that small sheep flocks could become at least marginally economical on small New England farms. It was an attractive proposition to many part-time hobby farmers and rural lifestyle folks, so we got into it.

Shepherding was enjoyable and educational, if only just barely profitable. First, we had to get our pastures fenced to keep sheep in and dogs out. At the Earles' we built a permanent five-wire, high-tensile electric fence around four acres of pasture, woods, and brush: a secure home base for the flock. On our summer pastures we lived dangerously, relying solely on portable electric fencing that we moved with the sheep. Portable fencing is made of polywire, plastic twine with stainless-steel filaments braided in among the nylon threads. We used a ready-made polywire mesh that comes in 150-foot rolls, with attached jab-in posts. With this "electronet" we could enclose a half-acre section of pasture in an hour or less, weaving our way around swimming pools, gazebos, flowerbeds, Japanese maples, and whatever other obstacle the suburbs threw in our way.

Confining sheep with polywire alone was risky. The tiny strands offer much greater electrical resistance than heavy-gauge high tensile wire, so the fence does not deliver such an arresting shock (trust me). The sheep usually respected the polywire, but once or twice every year they would break out and go wandering through the suburbs in search of a swank lawn. Faith and I would get a call from the police and go round them up, usually in the middle of the night. Luckily the sheep wore bells that Dr. Earle had bought, so they weren't hard to find. For the most part, neighbors were more amused than annoyed by these quaint trespasses on their property. In fact, many people were so delighted to have tinkling sheep grazing in their neighborhood that they helped us out, keeping an eye on the flock and even watering them for us while we were away.

A more serious threat than loose sheep was loose dogs. Weston was full of free-running dogs, and free dogs love to run sheep. The town did not have a leash law because freedom for a dog to run is strongly associated with "rural character" in the suburban mind. Suburbanites do not appreciate that in real rural areas the responsibility of owners to control their dogs is universally recognized, and that even much-loved, amiable dogs are summarily shot if they get into mischief with a neighbor's stock. Most of our suburban neighbors understood the problem and let us train their dogs to the fence by allowing them to be shocked under controlled conditions, so that they gave the fence a wide berth from then on. But every now and then a strange dog would come trotting along, see the sheep, charge the fence, and get in. In the decade or so we had the flock we lost two sheep to dogs. Friends of ours have suffered much worse depravations from both dogs and the bold eastern coyote, even behind high-tensile fences. When it comes to raiding domestic stock, Weston's coyotes seem to prefer cats, which are more plentiful than sheep.

Dogs are one drawback to raising sheep in the suburbs. Moving livestock among many small pastures is another. Faith and I learned a lot about how to handle sheep in our years as shepherds. They will usually come to grain or even to the rattle of grain in the scoop; but even the lure of grain will not overcome fear. Once fear is in the air, the place you want the flock to go generates an invisible force field from which galloping sheep rebound as if by magic. How high can a terrified sheep jump? I once saw a sheep's hooves pass through the space occupied a moment earlier by the head of a brawny high school football player, who had just thought better of tackling her and taken a dive. Six of us captured that sheep (whose name was Chocolate) some minutes later in the corner of a neighbor's tennis court. (If you ever lose your flock in the suburbs, chase them into a tennis court—they make very handy corrals.) We still have Chocolate's picture on the wall and her handsome pelt on one of our chairs. She was a good sheep but a bit bonkers.

There are subtle ways to outwit a sheep and get her to move in the direction you want, but subtlety is often lost on these creatures. A bright-eyed border collie is a much better bet. Now and then we enlisted the help of Ellen Raja and Betty Levin and their dogs from Lincoln and moved the flock right down the road. There was great crack in it, as they say back in Ireland. We would warn the police beforehand that we intended to stop traffic with a flock of sheep, which they found so amusing that they often dispatched a squad car or two to lend encouragement. Once we held up a funeral procession until the sheep were off the road, a scene more reminiscent of Galway than Weston. One thing that helps in running a community farm is a police department with a sense of humor: over the years Weston's police have been uncommonly tolerant and helpful through the many escapades of Green Power and Land's Sake.

Herding sheep this way was so much fun that Faith and I got a border collie pup of our own, which we named Jake. Jake is a good dog, but in spite of our best efforts to train him, he never really got the hang of herding sheep. He doesn't have the sharp border collie eye that sheep respect. A shepherd with a clever dog can walk serenely in the direction the sheep are to go or just whistle commands a mile across the glen to the wee streak of a dog who circles the flock, keeping everything in order while your man stands at his ease, looking natty. When Faith and I herded sheep, we made Jake stay in one spot and circled them ourselves.

Mostly, we moved the flock by trailer. In the early years we kept the sheep in a single large flock and moved them every few weeks. Later we broke the flock into smaller units and distributed them among several pastures for the season. We then grazed all these flocklets rotationally within their small pastures. This meant we drove or cycled around every evening to fill water tubs and move fence but saved on the number of times we had to put in an entire day hauling the whole flock across town. We kept the lactating mothers and lambs on the best grass and parked the dry ewes and yearlings on poorer scrub. After the lambs were weaned in midsummer, we switched them onto our lushest grass so they would keep gaining weight. Suburban pasture management is a complex business. As time went by, we became more adept at getting all the sheep where they needed to go, when they needed to be there.

Every fall we turned the ram in with the ewes to breed. Before putting him in, we strapped him into an elaborate harness that held a red crayon under his brisket. In this annual woolly fertility rite, I would spend about half an hour holding the big fellow and scratching his ears while Faith tried to remember which way the leather straps went. With his crayon snugly in place, every time he mounted a ewe he marked her on the rump, so we could note which ones he had covered when we came by to water the flock in the evening. Sometimes he left a big red smudge, and sometimes he left barely a streak. From the day of insemination we could calculate the day when each ewe was due to lamb in the spring. We timed the breeding so that as many lambs as possible would be born during February vacation, when Faith was out of school for a week. Early lambing meant the lambs would be just old enough to begin grazing with their mothers by the time the grass greened up in May. As in many of our farming projects, we made subtle compromises between the cycle of seasons and the cycle of the school year, and in this case it worked out just fine.

Because we always left the ram in for two menstrual cycles, and there were always a few ewes who did not settle until the second go-around, lambing dragged on for about six weeks—from mid-February until nearly the end of March. Generally speaking, older ewes lamb without difficulty, while first-time mothers are often oblivious to what is befalling them and don't know what to do about it when it does. Maternal instinct is not as ingrained as you might think; experience counts for a lot, even in sheep. Soon enough, both Faith and I learned how to extricate a lamb with a leg back, how to get a lamb breathing if the mother failed to clear its nostrils, how to strip the ewe's teats to start the milk and get a sluggish lamb up and nursing. Routine emergencies I could handle myself, but for hard cases like a backward lamb I always awakened my wife, who was born with a knack for such things—or just goes ahead anyway. We learned by doing.

Many sheep breeders let the lambing take care of itself, accepting a few inevitable losses in exchange for the savings in labor and sleep. Not Faith. Driven by her iron will, we checked the ewes every four hours throughout the birthing season, day and night, without fail. It is unusual for a ewe to give no sign of impending birth at one check and then to get into serious trouble and lose her lamb within the next four hours (although it does happen), so we were usually there in time to intervene when nature went awry. Because Faith was a teacher, Dr. Earle a doctor, and I a mere graduate student and farmer, I was assigned the late checks. Every spring I shifted my schedule back about six hours and got used to never being quite awake during the day. I came to enjoy this. It seemed to suit the time of year. I liked sitting in my study at the computer and mapping old farms or writing, the world outside quiet and peaceful—in fact, I began writing this book in those hours. I liked going out to the woodpile every few hours to see what the weather was doing, keeping the faithful woodstove packed full of oak through the coldest nights. I liked climbing into my insulated coveralls and boots and heading for the barn. The walk down the winding driveway through the woods and along the lane, the dark shape of the dog leading the way, became so familiar that I found my way more with my feet than my eyes. I didn't bother to carry a light. I liked the darkness, although in the weird orange glow of the city reflecting from the clouds there wasn't much darkness to enjoy except on perfectly clear nights. I liked being abroad at all hours, tending the flock by day or night. Like Thoreau, I became an inspector of snowstorms.

What I liked most about lambing season was the way the current of time was checked and turned in a long eddy, without any sense of progress except that the barn cellar was filling with lambs. The weather in Massachusetts gets warmer, on average, between the beginning of February and the end of March, but a cold, snowy week seems as natural at one end of this period as the other. The ground freezes, thaws, and freezes; the snow falls, melts, and falls again; the mud comes and goes; and for weeks on end the world seems full of rain and mist. At lambing time, the weather outside matched the foggy climate inside my head. Because I was as liable to be walking down to the barn to check for lambs at two in the morning as two on a bright afternoon, I did not seem to be moving forward, clicking off the days, but just drifting around and around in place. I was never so conscious of every degree of the earth's continuous rotation in and out of light and darkness, of cycles within cycles, as when I was lambing. As long as lambing continued I felt that the press of the impending growing season was held in abeyance. I needed only to attend to the demands of birth as they arose and not look too far ahead.

Usually when I arrived at the foot of the stairs in the barn cellar all was peaceful and unchanged in the cold and quiet, the ewes chewing their cuds, grinding their teeth, and grunting with the heaviness of their pregnancies. In the mixing pen a few lambs would be up, jumping on and off and treading in place on their mother's back, causing the ewes to crane their necks gratefully—sheep love to have their backs scratched. When new lambs had been born or were about to be, I would get the ewe into a lambing pen if she wasn't there already. Once the lambs were out and breathing and reasonably well cleaned and dried by their mother or by towel and heat lamp, I would go up to the hot water spigot on the outside of the Earles' house, mix a bucket of warm water and molasses, and bring the ewe a flake of good second-cut hay. Sheep who have just delivered are usually thirsty and ravenous and need to keep their blood sugar up. While the mother munched away, her drawn-in sides still heaving, I would trim the lambs' cords and dip them in iodine and try to latch the lambs onto their mother's teats if they didn't seem to be getting the idea on their own. An experienced mother might turn back from the hay to help out, licking them in the rear end to stimulate nursing. Lambs from experienced mothers seem to be born smart themselves. First-time lambers, on the other hand, often tried to kick their clueless, groping lambs off the sensitive, swollen udder. Getting that first drink of colostrum is crucial to a lamb, so a shepherd needs to see them up and sucking before he can go home to bed. Inducing a lamb to nurse on an inexperienced mother, guiding the sloppy little creature's jaw with one hand and wiggling it under the tail and holding it on its feet with the other, while pinning the flighty ewe to the wall with your shoulder, can be either a very satisfying or an excruciatingly frustrating business. There is a fine line between rushing nature and being too complacent with a confused pair of novices who will do poorly or perhaps die if you do not intervene. When everything was finally going smoothly, I would turn off the heat lamp and walk back up the hill, to write or to sleep.

By April a brat pack of black and white lambs was charging in and out of the barn and bouncing off the walls, skidding periodically to their dirty knees to pound away for milk beneath their mothers. Meanwhile, the ewes' fleeces were getting long and ready to come off. We sheared the ewes and yearlings in early May, before taking them to their summer pastures. Our shearing was done by Kevin Ford and Carol Markarian, the only professional blade shearers still working in New England, as far as I know. That is, they use old-fashioned hand-powered steel blades rather than electric clippers. Their rates were reasonable and their work impeccable. We looked forward to their visit every year, for the conversation and for all they could tell us about the condition of our flock as well as for the pleasure of watching them at work.

When shearing, Kevin and Carol bend double over the animals, controlling them mostly with their feet and knees. It hurts my back just to see them do it. The sheep are either seated on their rumps or curled over on one side, held in helpless positions in which they generally remain calm when handled by calm people. The blows with the blades that remove the fleece intact follow a precise order that has been honed over the centuries. The blades are sharpened to knife edges with almost no bevel, unlike the steep angle of scissors. They glide along just above the animal's skin, taking off the wool in smooth sweeps. Like any well-practiced art it all looks effortless, but I can assure you it is quite difficult. A few years ago I got a good pair of blades, ground them to a respectable edge so they cut nicely, and learned to use them to crutch the ewes before lambing. Crutching is the careful trimming away of filthy wool and tag ends from the hind legs and belly around the ewe's vulva and bag, which clears the way for delivery and for the lamb searching for the nipple. I got so I could crutch a ewe in five or ten minutes, about the time it took Kevin or Carol to shear an entire sheep.

Faith saved the fleeces with the best springy crimp and the nicest colors for handspinning. She spun yarn by the woodstove on winter evenings and at various fairs and demonstrations. She also sold a few fleeces to other handspinners. Most of the fleece we sent to a small commercial mill called the Green Mountain Spinnery in Vermont to be custom spun into yarn, some creamy white, some nearly black, and some a lustrous gray. The natural colors went beautifully alongside one another. Faith used this yarn and her own handspun for knitting and weaving, and we sold some at the farmstand and through a yarn shop in Weston Center.

Wool yarn is worth little money and doesn't sell very fast. Cheap synthetics have long since undercut wool as a leading commercial fiber. Getting top-quality fleece was Faith's motive for going into sheep, but we knew very well that wool would never pay the bills. Most of our cash income for the project came from freezer lamb and from tanned sheepskins. The grass-fed lamb was delicious and sold well. Customers signed up for a half or whole lamb at the farmstand every summer, and Faith distributed the meat as soon as we got it from the butcher in the fall. We salted the raw skins from the slaughtered animals and sent them to Quakertown, Pennsylvania, to be tanned. They came back light and fluffy and washable and sold well at the stand, netting us almost as much as the meat. After a few years, we had at least the rudiments of an economically viable enterprise.

We kept the best-looking ewe lambs to breed, and the flock began to grow. Faith didn't much desire a large flock, but I did. I wanted to see if it was practical to run a commercial sheep operation in the suburbs, utilizing the portable fence to graze small pastures. As long as I was reorganizing my life for two months to check at all hours for new lambs, I thought it would be worth it to have as many arrivals as we could house. The same logic held true for many of the daily chores—it couldn't be that much more work to feed and water fifty or even a hundred ewes than ten, once you were down at the barn anyway, I reasoned. We let the flock grow to about thirty-five breeding ewes and went looking for grass. Soon we had recruited field owners all over town to buy fence and entertain sheep on their land. I was seized with the idea of reviving pastures in Weston.

I became a man with a mission. Nothing brings rural character back to life like putting a few sheep in an old field. We were awakening a sleeping landscape. Weston was full of old stone walls running through deep woods and of brushy fields that had slowly contracted from large rectangles to small ovals as the forest crept in every year behind the circling mower. Like most New England towns, Weston was covered with hayfields and pastures until the land was abandoned to trees and subdivisions. I didn't want to see the landscape cleared of forest again, by any means, but there was great satisfaction in pushing the woods back over the walls here and there, uncovering some vestige of the Weston of our forebears.

The traditional New England landscape was formed by a pastoral economy that declined largely because ecological shortcomings degraded the land. That was one of the central arguments of my dissertation, so I was in fact conducting legitimate scholarly research as we went about restoring pastures for our sheep. On the whole, New England is well suited to grazing, but its soils and climate do present some special challenges—and New Englanders failed to meet them. I thought I had a grasp of where my pastoral predecessors had gone right and where they had gone badly astray. Would that insight and some flexible modern technology bring the traditional landscape back to life?

The Rise and Fall of Mixed Husbandry in New England

From the moment of English settlement until well into the middle of the nineteenth century, New England was primarily a pastoral economy, for the most part a cattle economy, although it had its sheep moment. Farming was diversified, but the stock drove the system. This was true not only of upcountry New England, but even of towns like Weston whose hills are within sight of Boston.

The long pastoral era preceded the era of commercial milk farming and market gardening that I discussed in chapter 2. Not until about 1850 did the streamlined system that provided such concentrated products as vegetables, fruits, poultry, eggs, and fluid milk to urban consumers begin to take shape. Before that, for most of two centuries farmers in towns like Weston practiced a form of mixed husbandry that combined livestock and tillage to supply the necessities of life, along with some surplus for cash sale. Mixed husbandry produced butter and cheese, salt beef and salt pork, corn and rye, leather and wool, cider and firewood all from local soil. This was the farming system that created the classic New England landscape, the world that Henry Thoreau lampooned at its height and that Robert Frost lamented in its decline. It was a world that deserves both to be celebrated and held to ecological account.

If cows drove farming, then it follows that the most important crop of all in New England, the foundation upon which almost all other production rested, was grass. Grass is the key to understanding the region's agricultural history. From the beginning, the fundamental problem in New England farming was how to grow enough grass to feed the cows. This challenge shaped and constrained what might be called the yeoman period of mixed husbandry in New England in the seventeenth and eighteenth centuries. When the demands of the market on the land greatly increased in the nineteenth century, the challenge of providing more grass threatened to break New England farming. This might be called the commercial period of mixed husbandry, which evolved into the era of concentrated products. As we consider what kind of agricultural system would be sustainable and desirable in New England in a more self-reliant future, we need to reexamine the grass-based mixed husbandry system that dominated the region during the long period when it *was* largely self-reliant. We need to understand its ecological virtues and flaws and to mark well the forces that finally brought it down.

When English farmers arrived in Massachusetts in the 1630s, it did not take them long to realize that raising cattle would be their main enterprise. The land reminded the immigrants of certain places from home: the Weald of Kent or the Pennine foothills of Derbyshire; upland districts where livestock had long dominated the economy and tillage was limited mainly to meeting local subsistence needs. The New England soil was too wet for the plow in some places, too stony in others, and sandy and dry where it was plowable. The topsoil was thin, sour, and, unless manured, easily exhausted of available nutrients. This land was never going to produce ready surpluses of grain, let alone staples like tobacco for export. Except in a few favored regions like the Connecticut River valley, wheat and barley did not thrive. The settlers soon adopted coarser rye and native maize as their everyday crops. These two grains made up the meal for the farm family's daily bread, unleavened hoecakes known as Johnnycake or rye'n'Injun, baked on the open hearth.

The wealth of the new country, such as it was, lay in cattle. Cows provided a good part of the year-round food supply directly through cheese, butter, and salt beef, along with whey to fatten hogs. Growing the corn and rye bread crops required cattle as well, both to plow and cultivate the arable land and to dress it with manure. Dung was essential to Yankee farming from the beginning. The New England forest had not created rich soils, and Native-American farmers had not worried about improving the soil in order to work the same fields year after year. They employed forest-fallow shifting cultivation methods and moved on to new ground after a few growing seasons. But the English farmers who appropriated these planting grounds had long relied on manure to fertilize their arable land. Their husbandry was organized around both providing subsistence and producing a surplus of cattle, by closely integrating grassland and tillage. They set out to combine Old World and New World crops and livestock into a workable system over a diverse and difficult terrain. Thus they established the agrarian pattern that would prevail in New England for more than two hundred years.

To run such a mixed husbandry system, the English settlers had to find grass to feed their cattle—not a trivial problem in a new land with hotter, drier summers and longer, colder winters than they had been accustomed to back home. The cattle required pasture for the summer and hay for the winter—lots and lots of grass. But New England as these English husbandmen encountered it was not a region of extensive grasslands. It was largely a forested region that had supported few native grazers to compare with the great herds of bison that ranged the prairies a thousand miles to the west. The native fauna of the eastern forest were primarily browsers such as white-tailed deer. If the native people used fire to maintain open places within this landscape (as many scholars believe they did), it was to support their wild stock on scrubby browse more than on grass. At first, the newcomers' cattle, too, had to pick up a living on the huckleberries and bracken that grew beneath pitch pine and oak. It was slim pickings and a hard living for cows. The only reliable grass to be found grew in wet places: salt marshes along the coast and open meadows along broad stretches of rivers and brooks inland. Places with large stores of native grass became the sites of the earliest towns in Massachusetts Bay colony, among them Watertown, Concord, Dedham, and Sudbury.[2]

Weston entered the world in the seventeenth century as Watertown Farms, a collection of privately owned, largely unsettled blocks of land at the western end of Watertown. Watertown had been incorporated in 1632, and most early settlement was clustered at the eastern end of town near tidewater on the Charles River. Many of the proprietors of Watertown soon granted themselves farms in the hills a few miles to the west. At the time, the term *farm* did not mean land that was devoted to cultivation—the word for that was *improved,* as opposed to *waste. Farm* distinguished a landholding that was consolidated in a large block rather than dispersed in many smaller pieces, as had been normal in common field husbandry. For the most part, the grantees of Watertown Farms banked this wasteland for the future settlement of their descendants, in the meantime using it for rough grazing. One farm block, located on the far edge of town against the Sudbury line, belonged to the town itself and was leased to a series of men who kept the community's "dry stock" there for the grazing season. This ancient "cowpen farm," long a place of barren cows and retired, fattening oxen, is now occupied by some of the finest homes in Weston.

Through most of the seventeenth century, the milk cows of Watertown were collected into several herds that ranged out daily in designated directions, grazing over both common land and privately owned wasteland that had been pooled as additional commons. Most New England towns handled their stock in this way during the first decades of settlement, as yeomen had in many parts of England time out of mind. Sometimes Watertown had three common herds and sometimes four, as the selectmen labored to adjust the town's increasing grazing needs to its lagging resources. Fines were levied against those who did not herd their cattle properly, letting them run loose and merely sending their children after the homeward-wandering cows in the evening to make a pretense of herding. Ten acres of pasture were required for each cow, oxen, or horse, and two acres for each sheep that was to be grazed. The numbers give an idea of how rough the grazing was; today, just one acre of good New England pasture will support one cow or five sheep for the summer, a tenfold improvement. The system of common grazing lasted through the late 1600s, as long as the bulk of the town's backland remained unsettled.[3]

Watertown's sheep, prey to wolves and requiring better grass than other stock, were herded separately, on commons closer to the village. It was a rare and endangered sheep that wandered as far as Weston in the seventeenth century. By town order rams were removed from the flock from July 1 until October 25—doubtless to prevent births between December and the end of March, when little quality hay would be available for the lactating mothers and no fresh feed was in the offing. Unlike sheep, swine—with high reproductive rates, omnivorous diet, and fierceness of mien—could fend for themselves in the woods. Some towns herded swine in "hogpen walks" tucked in some rocky back corner where they could fatten on acorns, but the evidence is that many hogs wandered loose through the woods and raided the planting grounds, generally wreaking havoc.

By the last decades of the seventeenth century the outlying corners of Watertown were being settled. All around Massachusetts Bay the last unenclosed commons were divided among private owners, and new communities were beginning to hive off from the "ancient" mother towns. Watertown Farms established a parish church in 1698 and was incorporated as the town of Weston in 1713. Common herding faded as resident husbandmen improved pastures on their land. They erected rail fences and stone walls around their grazing lots to keep the stock in, as they had earlier enclosed their tillage land to keep the stock out. They cut trees, laid out driftways, or paths for the cattle to follow to pasture, and dug out and stoned in springs for watering places. In short, they remade the land into working farms in the modern sense of the word.

As the land was being grazed and fenced, important ecological changes were taking place underfoot. As pasturage, the grazing lands were slowly improving in quality. As woodlands they were, by the same token, steadily declining. The combination of grazing and trees on the same land has seldom worked to the advantage of trees. The forest was inexorably diminished, sometimes gradually as grazing stock nibbled back hardwood regeneration, sometimes instantaneously as trees were cut and the slash burned. This created an opening for European cool-season, sod-forming grasses such as bluegrass, fescue, crabgrass, and bent—species well suited to New England's climate and long adapted to frequent gnawing. These grasses booked Atlantic passage with the Puritans and their beasts, spread their seeds in the dung of grazing livestock, and took root in the New World. White clover also sprang up and throve in the nitrogen-poor soil, especially where fire had contributed a layer of sweetening ash. The first decades after pastures were enclosed were good ones for grazing—far more productive for stock, at any rate, than the forest and scrub that had greeted the first generation of settlers. Throughout most of the colonial period the pastures of Weston proliferated, and grazing capacity was on the rise.

As the first generations of English immigrants toiled to get their pastures in order, they faced a second, parallel challenge: where to find sufficient fodder to bring their cattle through the long, tough New England winters. They had not been used to such rigors of cold in England, where no more than a month or two lacked some grazing. In New England the hungry gap stretched to half the year, from November through April. In England, sufficient hay had been cut in coastal fens or in a few choice, well-watered meadows along streams, and so the settlers in America turned first to the same familiar sources. They did not have any experience growing domesticated hay that included legumes like red clover and alfalfa—that innovation swept England just after the Puritans left for America. The legume revolution so critical to the development of agriculture in Europe took more than a century to cross the Atlantic, in part because in New England the native meadows were plentiful and proved adequate for several generations.

In colonial America (as in England) the term *meadow* meant not just any expanse of grass, but a low-lying, wet one. These native meadows, whose grasses often mixed with harsh sedges or woody stuff like buttonbush and black willow, also required several generations of sweat to be brought to a fair state of production. The farmers had to cut out trees and brush in swampy sections, dig intricate systems of interconnecting drainage ditches, and painstakingly keep them clean. Drainage made the meadows passable to oxcarts that carried off the hay and encouraged the more desirable grasses to thrive in place of the unpalatable marsh and swamp species that dominated permanently flooded land. By the end of the colonial period, most of the wetlands in towns like Weston had been rendered into serviceable mowing ground. The meadows ranged from small pockets to sweeping expanses of native species like reed canary grass, cordgrass, bluejoint, fowl meadow grass, and redtop. Not the best by today's standards, this meadow hay was coarse and low in protein but reliable and filling, and by all accounts the thrifty red Devon cattle throve on it. Agriculture in colonial New England was built on meadow hay, which the nineteenth century denigrated and the twentieth century has all but forgotten.[4]

Native hay meadows allowed farmers in towns like Weston to keep large stores of cattle, and for several generations, by and large these farmers prospered. Few became wealthy but most raised large, healthy families. Many passed along ample landholdings to their children to improve and often lived to see their grandchildren come of age around them. They fed these families largely on what the cattle provided, directly or indirectly. As the meadows supplied hay to the cattle, they also supplied nutrients to replenish the tilled fields in the form of manure. Farmers in colonial New England spent weeks every spring carting dung from their barnyards to their cornfields—the charge that American farmers wasted manure may be just in other regions and other times, but not here. Farmers in this district were never blessed with the kind of soil that could have endured such neglect. The evidence shows that they tilled the same fields year after year for generations, which indicates they manured it faithfully. Behind that manure stood the hay meadows.

The saving grace of this system of husbandry was that the low-lying meadows were next to inexhaustible. They occupied deep alluvial soils that had been accumulating for millennia and that continued to be replenished by sediments that trickled down from the uplands to be deposited by annual winter floods. In fact, many husbandmen deliberately "flowed" their mowing land during the winter, an ancient English practice. As long as farmers in

Massachusetts towns kept their herds and their cultivated acreage within the limits of what their meadows could supply with hay and manure and did not overcut their woodlands, they were on reasonably secure ecological footing. They were never going to get rich, but getting rich was not high among their aspirations—for the most part, these yeomen strove for what they called a competency and a respected place in their family and community.[5]

An eighteenth-century New England farm included several kinds of land, each with its own function, integrated into a working ecological whole. Although these farms were more compact than they might have been under the old commons system, many were still strikingly dispersed. The land making up each farm seldom lay in one contiguous homestead but was spread in pieces over the surrounding landscape so as to take advantage of the diversity of available resources. Tillage tended to be the most consolidated close to home because of the heavy labor of carting manure from the barnyard. Typically, there were multiple pastures, one near at hand and a few others as far as a mile away. Many farmers owned large tracts of rough summer grazing for their dry stock on land they had acquired in new backcountry towns farther west and north—land destined to become new farms for sons and grandsons. There might be a hay meadow in wetlands within the homestead and several more mowing lots up to three miles away, totaling ten to twenty acres to be cut and carted home every summer. These meadows also provided fall grazing on the aftermath, or second growth, after mowing. An old orchard might stand on an acre or two behind the garden, and a younger orchard on a marginal hillside up the road. Finally, most farmers owned several woodlots squirreled away on rugged uplands and ragged swamps, often a mile or two from home, to which they would repair in the winter to cut firewood and timber as needed.

Early American farms have often been portrayed as primitive and ecologically awkward at best. At many times and places, especially during the helter-skelter nineteenth-century conquest of the continent, this notoriety may be deserved. But there was little that was random or thoughtless about the organization of New England farms. They marked an admirable ecological adaptation of the English mixed husbandry system to a new environment. Both native corn and English rye were well suited to the warm, sandy tillage soils and could withstand a hot, dry summer once established. Squashes and beans had also been adopted from Native Americans, contributing the famous pumpkin pies and baked beans to the New England diet. Apples had proven ecologically superior to barley as the principal beverage crop (see chapter 4), and hillside orchards flanked the roads. European cool-season grasses had taken hold in upland pastures, while native meadow grasses were a reliable source of hay in wet lowlands. The forest, as yet not severely depleted, met a wide range of needs. Cattle bound the system together, affording milk and meat, traction and transport, leather, manure, and even light—linking each part of the farm to the next and the landscape to the barnyard and hearth. Land and community formed a complex, diverse, tightly woven web.[6]

The residents of towns like Weston were surrounded by neighbors and kin, people with whom they were enmeshed in networks of local exchange. While they cooperated closely with one another out of necessity and obligation, they also competed closely for the ever-tightening supply of land. Buying land required cash, as did purchasing a few imported goods such as coffee, tea, sugar, metalware, and fine cloth. Many families supplemented their income by engaging in a trade—blacksmithing, tanning, cooperage—thus supplying special goods to the local economy that not every family could easily manufacture at home. Some products were sold for hard cash, but most were traded on account—simply credited and debited in the books farmers and tradesmen used to keep track of dealings with their neighbors. Probated estates frequently record the petty debts of a community living in each others' pockets: a ton of meadow hay, a half-day of a boy with a team of oxen fitting land for planting, a cord of oak fuelwood, a day setting out apple seedlings, a cartload of manure. The tangled web of accounts sometimes spun along for generations.

Men and women participated in largely separate exchange networks. Men's outdoor work slackened somewhat in the short days of winter. Inside the home, women toiled without respite to transform the raw products of the land into usable food and clothing and formed their own neighborly circles by exchanging garden produce, spun yarn, woven cloth, and midwifery services. Older daughters might go to work for a few years as servants with neighbors who needed domestic help. Not many households were self-sufficient or aimed to be, but the community as a whole produced the great bulk of what it consumed directly from the local environment.[7]

The most notable flaw in the ecological fitness of these New England yeomen was not agronomic, but demographic: there were too many children. The most glaring social flaw was that women were confined to lives of low status and grinding toil bearing, feeding, and clothing the multitudes of desired offspring. Large families were necessary to this patriarchal way of life, and in the healthy New England climate large families were produced. The making of a yeoman's success was many sons to help him work the land when he was in his prime. The mark of a yeoman's success was to see those sons settled around him on enough land to make a steady living and taking a respected place of their own in the community. But as mother towns like Watertown split into daughter towns like Weston that filled in turn, finding enough nearby land for more than a single heir became increasingly difficult for many families. By the middle of the eighteenth century, with the Indian wars behind them and the frontier relatively quiet, fourth- and fifth-generation Massachusetts husbandmen began streaming out to settle new towns in the center of the state and in Vermont, New Hampshire, and Maine, beginning the process of farm making all over again.[8]

These emigrants did not leave behind towns in which poor farming methods had exhausted all the available land. All the trees were not yet stripped from the hometown hills. Before that point was reached, these communities ran up against the limit of the key resource in their agricultural system, meadow hay. When a community is obliged to make balanced use of diverse local resources, its growth may be at least partially checked before damaging the landscape as a whole. As meadow hay became scarce, the expanding yeoman economy hit a snag. There was still room for agriculture to grow within the old towns, but not enough fuel (in the form of hay) to feed its growth. More upland could have been cleared from the forest to pasture more cattle—but there was not enough hay to feed additional animals through the long winter, so they did not appear. More cropland could have been cleared to grow more subsistence bread grains—but in the absence of more cattle to manure this cropland, the total yield of the corn crop could not be greatly increased. In fact, in many towns per-acre yields began to drop because of the scarcity of manure. With its much lower population density than Europe, Massachusetts was hardly on the brink of starvation, but it was becoming difficult for young husbandmen to step into the ranks of the yeomanry and attain the rude but comfortable station in life that their fathers and grandfathers had enjoyed.

By the end of the colonial period, rural New England was at a turning point. The mixed husbandry system worked well, but it had reached limits that demanded cultural, economic, and ecological adjustments. The response to these limits in turn played a role in the demise of the world of locally self-reliant communities. The age of the yeoman was ending.

The decades following the Revolution saw broad cultural changes in the new nation. Americans began to strive less for the comfortable subsistence and filial respect achieved by playing a preordained role within a large family and a covenanted community and more for individual success as measured by material prosperity. New Englanders took a leading role in this transformation, and the road to greater prosperity (at least for those who succeeded) ran increasingly to market. Farmers had always marketed part of their crop, but now they began to look deliberately for those crops that brought the best cash return. In the process they ceased being yeomen and became commercial farmers. By specializing in a few cash crops, farmers in towns like Weston were able to break through the limitations imposed by the local subsistence economy they had created and extract more wealth from their land—at least for a while. These increases in productivity have been saluted by many economic historians, but I believe the evidence shows that while some improvements were genuine, taken as a whole the market transformation put rural America on the road to ruin. In the course of this commercial revolution in Massachusetts many farmers were bankrupted, and within a few decades the land itself was exhausted. The process continues today, but on a global scale.

What these New England farmers undertook was to substitute commodities imported from other regions for many of the subsistence goods they had been producing for themselves. This allowed them to concentrate on growing products for which they had a comparative advantage, to earn cash to buy the rest. They sold their produce to their Yankee cousins, who were replacing artisan shops with factories in nearby mill villages and factory towns and no longer feeding themselves. This change in the scale of production and commerce, along with a flourishing maritime trade, the southern cotton boom, improving internal transportation, and the application of water-powered machinery to manufacturing, was the key to economic growth in the early American Republic. One crucial factor for the nation as a whole was the crossing of the Appalachian mountains, which gave access to a vastly expanded storehouse of ready natural resources in the continental heartland beyond. But New England farmers were in the thick of it, finding fresh agricultural resources right under their noses, within their familiar landscape.

We can follow this change in commodity after commodity. Soon after the Revolution, farm families in New England cut back on growing and eating their own coarse corn and rye flatbread and began baking not just the "upper crust" of their pies, but their daily bread with wheat flour shipped in from Pennsylvania, New York, and the West. They continued to grow corn to feed their cattle, but in place of rye they grew oats—not to eat themselves but to sell for horsefeed, fueling the rapidly growing transportation sector of the day. Even before the building of the railroads commerce was picking up speed, driven by the horse. New turnpikes and bridges were being constructed left and right. Where lumbering oxcarts had dominated the roads, horse-drawn stage coaches and wagons began passing them by.

Within the houses by the roadside changes of equal magnitude were taking place. By the 1820s, women largely ceased producing homespun linsey-woolsy cloth and turned with relief to fabric purchased from the proliferating textile mills. Whereas once women spun yarn by tallow candles, now they might embroider finished cloth by whale-oil lamp. The spinning wheel was put away in the attic. Even light was increasingly purchased: it had its source in the oil of sperm whales harpooned in the Pacific Ocean, stripped of their blubber, boiled down, and brought around the world to Nantucket and New Bedford. What people drank changed as well. So did the way they heated and cooked. By the 1830s homemade apple cider gave way to tea and coffee as the principal beverages, and coal from Pennsylvania began replacing local firewood as the principal fuel. Iron cooking and heating stoves replaced fireplaces. The house itself changed—stud-wall construction from pine lumber milled in Maine began to replace timber frames hewn of Weston oak and chestnut. By the 1850s, rural people looked to the market for a large part of what they had formerly grown or made locally for themselves.

In order to take part in this expanding market economy Weston farmers had to find something of their own to sell. They concentrated on grass and livestock. Now the classic New England barns that still dot the landscape were built, to store more hay and house more cattle. The augmented hay crop was not only fed on the farm, but also sold to feed cows and horses in nearby cities and mill towns. As for livestock products, meat remained important until midcentury but paled in comparison to the rise of wool and milk. Newly settled towns in western Massachusetts, New Hampshire, and Vermont, where land was relatively cheap, became the scene of the early-nineteenth-century sheep boom that lingered through the Civil War. Hillsides were rapidly cleared of forest to pasture legions of Merinos, supplying wool to the region's burgeoning mills. Meanwhile, older towns in eastern Massachusetts saw a sharp increase in the production of butter and cheese for consumers in the rapidly growing cities. By the 1840s, fresh milk was being shipped daily into Boston along the early rail lines, one of which ran through Weston.

But where did farmers find the extra grass to make this possible? Producing more hay and livestock for market required an ecological as well as an economic revolution in New England agriculture. To keep more productive cows and sheep, farmers had to overcome the barrier imposed by their native hay meadows. They made the breakthrough by planting domestic "English" hay, a mixture of timothy, redtop, and red clover, mostly on newly cleared *upland* fields. A hilly town with few meadows, Weston was one of the only towns in Middlesex County in which upland hay production equaled meadow hay by the end of the colonial period. By the middle of the nineteenth century there had been little change in meadow hay, but twice as much domestic hay was being grown. New walls ran over the uplands as hayfields and pastures swept the remaining forest from the hills. Long-slumbering stones were rousted from the beds in which the glacier had rolled them and sledded to the field edges to make walls. During the first half of the century, Weston's woodlands shrank until they comprised only 10 percent of the landscape.

Meanwhile, the old reliable meadows themselves were also transformed. During the 1840s and 1850s, tile drains were laid down in many lowlands in Massachusetts towns, lowering the water table sufficiently to allow more profitable English hay to be planted in place of the coarse native species. Generations of yeomen had labored to put the basic drainage system in place. In the nineteenth century a deeper pass was made at "thorough drainage," as Yankee farmers took advantage of another new element in the New England landscape, the Irish laborer with his spade. A sharp rise in hired labor was one more sign of an increasingly commercial approach to agriculture. By laying tiles in the meadows and making a great leap over the uplands, farmers found that both the quantity and quality of their hay could be greatly improved. Grain yields rose, in turn, as more cows produced more manure to fertilize the tilled cropland. As agriculture commercialized, the landscape was transformed to a higher state of cultivation.

Or so it seemed for a time. But as the middle part of the nineteenth century wore on, problems began to surface. The rapid disappearance of the forest, for one, caused widespread alarm (see chapter 5). A problem of more immediate practical concern to many farmers was the decline in yield of their new upland hayfields and pastures, after only a few years of mowing or a few decades of grazing. In order to meet the demands of the booming market, farmers had cleared more land than they could maintain. As a result, they ran it down almost as fast as they cleared it. In particular, unsupported expansion drained the fertility from the grasslands, the main source of nutrients for the rest of the farm. One farmer in a central Massachusetts town called this all-too-common practice "skinning the land."[9]

A stingy but sustainable system had been transformed into a profitable but extractive one. Hay was still fed to cattle, and the resulting manure was still carted to the tilled fields to supply nutrients to the corn or potato crop. It had always been rare for manure to be spread back on mowing land—this was normally a one-way trip with no direct return. The flow of nutrients could be kept up indefinitely when it originated in the native hay meadows that occupied deep alluvial soils, providentially replenished by floods. The new English hayfields on the uplands were another matter. Whereas the old meadows sat at the bottom of the landscape and captured leaky nutrients to be recycled, the upland hayfields flushed them from the top all the more rapidly. Initial yields were good, but repeated cutting steadily removed minerals without making compensation. Liming was spotty at best, red clover did not thrive for long in the souring soil, and so nitrogen levels suffered as well. Barring some fundamental reform in management, these upland hayfields were on a slide toward exhaustion as steep and unequivocal as the back slopes down which their embedded stones had been discarded. The degradation of nineteenth-century New England farmland was not so much a legacy of extensive colonial practices as the abrupt consequence of the new commercial age.

Hayfields were typically cut until their yield dropped so low it was no longer worth the trouble; then they were relegated to ragged pasture. But things didn't improve much once mowing ceased. Grazing livestock did not deplete the soil as fast as haycarts did, but the trend continued relentlessly down. The best years of a typical New England pasture did not amount to more than a few decades. The popular idea that livestock fertilize land while they graze it is charming but naive—as a rule, a grazing animal returns only about three-quarters of what it eats to the soil behind it and walks away with the rest in the form of meat, milk, or wool. So Massachusetts pasture soils continued to decline in fertility and relapsed toward their acid native state as the cations departed for the city, bound up in flesh and bone. Without copious lime, perennial white clover growing in the pasture faced the same sour fate as its cousin red clover in the hayfield; and without flourishing clover there was no adequate source of nitrogen anywhere in the upland system. Farmers who could seldom afford to dress their mowing land had still less dung to spare for their pastures, nor often the wherewithal to cart and spread enough ashes and bones to replenish the depleted potassium and phosphate. Over the course of the nineteenth century pasture stocking rates fell steadily across the state. Even in towns where tax returns suggest that livestock and pasture remained stable, as in Weston, on closer inspection one finds that farmers were continuously clearing new pastures even as they let old, worn-out land return to forest.

As pastures were losing their nutrients, they were being mismanaged in the way they were grazed. New England pastures were not so much over-grazed, as badly grazed. New England has no tradition of superb pasture management to boast of. Tax figures show that nineteenth-century pastures were very lightly stocked, and as the century wore on they carried more brush and fewer cows. This is a clear indication of *continuous grazing*—the normal practice in this country until very recently indeed. Livestock were left in the same pasture for weeks or months or even the entire season. Under this dispensation the cows were free to eat whatever they liked, so they returned to the sweet, palatable plants as fast as they could resprout and let the coarse stuff grow up unmolested all around. Livestock picked out and exhausted the tender forage in much the same way that Land's Sake exhausts witchgrass in its pumpkin field, by beating on the foliage without remorse. After several years of such abuse, the best pasture grasses and legumes were all but extirpated. The land within pasture walls, so painstakingly cleared of trees and stones, was recaptured by meadowsweet and steeplebush, blueberries and huckleberries, juniper, cedar, and legions of young pines.

The burst of increased farm production that followed the market revolution in early-nineteenth-century New England was patently unsustainable. Growth rode on the back of upland hayfields and pastures that were quickly degraded. After a few decades of skinning the land in this way, the market ran out of fresh land to skin. The more pinched but more durable integration of ecological elements that had marked the locally self-reliant husbandry of the colonial period was gone. Not that the colonial system was perfect—there were better methods (for example, well-limed legume rotations) by which a genuinely more productive agriculture could have been sustained, but that is not what occurred. As is often the case when capitalism takes off, the most readily available resources were quickly extracted and consumed before people paused to consider what would happen next. By midcentury, farmers in the Weston area had little woodland left to "improve" for farmland and found themselves surrounded instead by the scrubby pastures they had actively unimproved.

Fortunately (no thanks to these farmers), the land was not ruined for the long term. The land had not offered a great treasure of nutrients in its topsoil to begin with and had not been badly eroded physically. The farmers were simply unable to sustain the *improvement* in the soil that was necessary for it to remain productive grassland, and so it relapsed toward forest. Enough trees survived in woodlots and fencerows to reclaim the land as fast as the pastures failed. As every student of New England history knows, that is exactly what happened next: after the Civil War, thousands of exhausted pastures began to grow up in the dense stands of white pine that still dominate the landscape.

But this sweeping return of forest did not signal the collapse of the rural economy. Far from it. As noted in the previous chapter, late-nineteenth-century New England farmers profited further by turning to concentrated products. They took another rational step: they abandoned their unproductive pastures and substituted cheap grain from the Midwest to feed their dairy cows, greatly boosting milk production in the process. Cheap western grain didn't destroy New England agriculture, it saved it. In Weston, between 1855 and 1895 milk production rose sixfold from 93,000 to nearly 600,000 gallons, even as pasture shrank from 2,500 to 1,600 acres.[10] Having extracted the limited soil nutrients from the thin topsoil of their stony uplands, New England farmers turned to nutrients from the fresher, deeper, and infinitely broader prairie topsoils being rapidly mined out west. Instead of learning how to go sustainably to the subsoil with lime and clover to rebuild their home ground, they went to the railway depot to pick up sacks of corn from Iowa and to send their milk cans off to the city. Such a degree of specialized production effectively marked the end of the old locally based mixed husbandry system, with all its virtues and faults. The cows may as well have been standing in a feedlot in the Midwest, as they increasingly are today.

This is the reforested world people in New England and indeed throughout much of the Northeast have inherited, although the details vary from region to region. This world resulted from a complex chain of events. Colonial farmers cleared more than half of the land in their towns to put in place a diverse mixed husbandry system. Their descendants cleared most of the remaining forestland to make way for commercial livestock production. But the upland hayfields and pastures that were cleared of stones returned to forest almost as fast as they were walled in. They were replaced by a more linear, streamlined system that ran largely on midwestern feedgrain. This ushered in factory farming with its inputs and outputs and shut the door on local ecological responsibility for the way we eat.

In chapter 2, I argued that the industrial system of agriculture that now brings virtually all of our food from outside the region is unsustainable because it depends on fossil fuel, fossil water, and unacceptable damage to the environment and to human health. I argued for the revival, in an improved form, of the era of concentrated products that prevailed at the turn of the century, by which we would again produce the bulk of our fresh produce, fruit, poultry, and milk on a smaller scale, closer to home. But that era of commercial dairy farms and market gardens was itself the heir of a failed pasture and hay farming system and marked a loss of home ground and a growing dependence on damaging practices elsewhere. Unfortunately, unsustainable farming is not a recent human folly, even though we may be going at it these days with unprecedented sophistication and scope.

The stone walls hidden beneath the trees on Massachusetts hillsides are a map to the past and perhaps to the future. When we look at the land reclothed in forest and ask what kind of agriculture will safely work here, we discover a more troubling picture of what has driven agricultural development than a simple abuse of modern technology. We are forced to look at deeper issues of soil depletion and erosion that have long plagued commercial farming in America.

From the beginning of European settlement, American farmers have exposed the land to persistent abuse by extractive practices. Such usage has often been excused as simply the rational response of a small population of pioneers to a vast patrimony of virgin land, a response that enabled them to shrug off laborious intensive cultivation in favor of a more economical extensive approach. The idea is often expressed in economic shorthand as a "high land-to-labor ratio" in early America as compared to Europe. I believe this neat formula oversimplifies the issue. By focusing on the rational behavior of individuals operating within a fundamentally irrational economic framework, it makes what was inexcusably stupid appear smart. In the first place, it blithely assumes the existence of so-called virgin land that in actual fact had to be wrested from Native-American people who were already deeply attached to it. The land wasn't just waiting there empty, it had to be taken, and that is instructive. Native treatment of the land may not have been ecologically perfect, but, having a relatively light population by European standards, they had not just naturally abused the land to the degree the invaders did. Obviously, something changed besides the ratio of population to land.

The idea that it is only natural to treat land by purely market calculus also discounts the tenacity of traditional European farming methods that embodied another point of view. Before being overthrown by a revolution in economic thinking, these traditions were carried to this continent along with strong cultural ideals of family and community that many European settlers deeply valued. People have made ecological mistakes in many cultures, but they don't just naturally abuse land at the rate achieved by American farmers unless something powerful is driving them on. What it really took to get extractive farming going in the New World was the advent of a full-blown capitalist approach to the land and its commodities, as practiced by a culture that was freeing itself from worrying too much about the future of the local places it inhabited. Call that rational if you like.

In the tobacco colonies by the Chesapeake, where a much weaker community structure was planted than in New England and the land was cropped harder for export, soil depletion and erosion emerged as problems within a century of settlement. During the nineteenth century this extractive plantation pattern swept across the Cotton Kingdom of the Deep South, where it prevailed well into the twentieth century. Much of this cottoned-out region was finally bought up by large timber companies and is now growing pines in monoculture plantations, so in a sense not much has changed—although yellow pine is admittedly an improvement over yawning gullies. In the northern colonies (and by the nineteenth century in the upper South as well), market extraction was tempered by mixed husbandry systems that included some crop rotations and kept some manure on the farm. Next to the Amish, colonial New England was perhaps the most notable exception to the rule of straighforward extraction. In this rocky corner of America, soil resources were severely limited, production for local consumption dominated, and market farming was initially underdeveloped. For a time some semblance of a balanced agroecological system took hold, as we have seen. Stony soil builds character—or so New Englanders once believed in their hearts and proclaimed to a skeptical nation.

The situation changed rapidly for the worse with the rise of commercial farming in early-nineteenth-century New England. Although a few farmers became more prosperous and did improve certain of their farming methods (better manure handling, for example), the rapid advance in production came at the expense of the equally rapid depletion of upland hay and pasture soils. Historians have not looked seriously enough at this process but assumed that most of this land was inherently marginal and usable only until better soils were unlocked out west. But marginal is as marginal does. More lime, more legumes, and better pasture management might have made possible a more sustainable improvement in the productivity of Massachusetts farmland. Instead, farmers followed the path of least resistance and simply ran out the immediately available soil resources. The result of economically rational expansion was a degraded landscape, leading to an equally rapid ecological contraction in farming.

Pressure was building to reform the upland hay and pasture system as the limits of available land were reached and New England farmers faced declining grassland yields. At the crucial juncture, however, a cheaper solution presented itself in the form of feedgrain from the Midwest. Grain could be grown more cheaply in the West precisely because the soils there were deep and fresh and not yet in need of manure. Commercial farming in essence cast off its faithful eastern wife and ran away to the frontier. There, virgin prairie soils suitable for mechanical cultivation and harvest on a large scale were to be had for the taking. Spurred on by railroad corporations that had land to sell and needed freight to haul, grain farming in the Midwest was established on an extractive commercial basis from the beginning. New England farmers discovered that, given the opportunity to buy feed corn for their cows for not much more than it would have cost to grow it themselves, they could realize a larger comparative advantage by using their best soils to grow more profitable fresh produce and abandoning their worn-out pastures altogether.[11]

Easy grain gave New England's exhausted pastureland some relief, but it was no solution for the farmland of the nation as a whole. Soil degradation was merely shifted from the eastern seaboard to the heartland and greatly accelerated. The rich, loamy mollisols that had formed under the prairie grasses had far more fertility to lose than the leached spodosols underlying the New England forest. Within a matter of decades midwestern farms were showing signs of serious nutrient depletion and pervasive erosion. The flood of cheap grain that flowed from the American breadbasket to provision the East Coast (and indeed much of the burgeoning population of industrializing Europe as well) culminated in the Dust Bowl. After the "Dirty Thirties," erosion slowed somewhat, but it did not stop. What goes down by water unnoticed is often more damaging than what goes up in the wind. Industrial agriculture has led to the loss, by some estimates, of half of the Midwest's matchless topsoil within a century of settlement. And it has enriched relatively few farmers.[12]

The problem is still not solved. The production of food in America continues to undermine the long-term capacity of some of the world's best farmland, in spite of the application of chemical fertilizers that replace the extracted nutrients but do not replenish organic matter or maintain soil structure. Rates of erosion have gone up and down in America during the past century, but they remain above replacement levels over much of the most productive soil. During times of crop surplus and low prices, such as the 1930s, 1950s, and 1980s, some of the nation's most erodible land has been taken out of production and soil conservation measures grudgingly implemented, at great expense to the taxpayer. Then, typically, when demand for grain rises and prices go back up, conservation measures are cast aside, and erosion resumes its accustomed pace. Across much of the Midwest, the loss of soil was as bad during the grain boom of the 1970s as it was in the Dust Bowl years.

Only a foolish optimist would believe that this descending spiral will not repeat itself, that we somehow know better in the 1990s as the stage seems set for another surge in world grain demand. Today, the old, cherished freedom to farm (destructively) is rearing its ugly head once again. Meanwhile, through good times and bad the quality of the soil, its organic matter content, texture, and microbial life, continues to suffer under continuous tillage and chemical applications. If we had a system of accounting that properly amortized the costs of repairing this damaged soil, we might realize the deficit we are running up by growing food on the cheap. Possibly none of the hidden costs of industrial agriculture discussed in the previous chapter, including oil dependency and environmental pollution, will ultimately be as debilitating as that most ancient of agricultural scourges, lost soil.

Such a realization should help concentrate northeastern minds upon the challenge that faces us in producing more food within our own region. If we are to return production in New England to anything like the level it once attained, without relying on massive subsidies of oil (which have their own pitfalls, as detailed previously), we will need to be equally careful not to fall back on ruinous subsidies of soil. New England even a century ago required large imports of meat and grain from the Midwest for direct human consumption and even larger imports of feedgrain to supply dairy and poultry production, to fuel workhorses, and indirectly to fertilize truck farms. To the extent that New Englanders continue to rely on such imports (which will inevitably be considerable given a large population and limited prime tillage land), midwesterners will need to be certain that they are being sustainably supplied. And to the extent that we revive a hay, pasture, and grain farm economy to provide at least some of our own staples, for example, by restoring a dairy industry on our own ground or by doing a larger part of our farmwork with horses, New Englanders will need to learn how to become better grass farmers than were our forebears. What can we learn from their mistakes? What would ecologically sound pasture and hay farming in Massachusetts look like?

Ultimately, some grass, livestock, and grain farming is a necessary and desirable part of sustainable farming in New England. Making the case for cows, however, is not so easy as making the case for sweet corn on the one hand or for cordwood on the other. Reviving the pastoral component of our landscape may require us to clear part of the recovered forest—not an environmentally popular idea at first glance. But there are good reasons to return pastures and livestock to at least some New England hillsides. Our climate and soils, properly managed, are well suited to growing grass, and there are livestock products we can produce economically within the region in this way. Livestock and grass can play a vital role in nutrient cycles that support market gardening and in supplying draft animals to work efficiently both on truck farms and in the forest. Finally, hillside pastures, stone walls, grazing beasts, and graceful barns are a pleasing prospect and offer abundant opportunities for rural residents to make healthy connections with the land they inhabit. Community farms have a useful role to play in promoting this revival. And communities have a strong vested interest in holding protective easements on a substantial portion of the open land underlying even private farms. The restoration of active farmland should be a matter of broad community values, not narrow economics. Left to its own sharp logic, the market will always skin the land.

Principles of Sustainable Mixed Husbandry

ECOLOGY

In spite of the region's bad reputation, New England has many soils with good potential for pasture. The climate is nearly as well suited to cool-season grasses as Northwestern Europe. New England has higher rainfall and cooler summers than most of the rest of the nation—someday, when American culture finally outgrows its agricultural adolescence and adopts mature husbandry, this is going to be an advantage in raising livestock. In the course of moving Land's Sake's sheep among small pastures scattered throughout Weston, Faith and I had ample opportunity to observe how different soils responded to grazing in both wet years and drought. We also (out of necessity) conducted many experiments with different grazing regimes, with seeding and reseeding grass and legume swards, and with clearing new pastureland. After a decade I have only a beginner's knowledge of this subject and no formula for success, but I can suggest some general rules for reviving a sustainable pasture economy. The first is to put the pastures on the proper soils. The second is to establish a good mix of forage plants that include a vigorous legume and to maintain optimum conditions for pasture growth with periodic liming. The third is to practice rotational grazing to keep pastures productive.

A good pasture soil in New England must be well watered—it does not need to be tillable land. You can have a decent lawn (which is nothing but a fancy pasture sod) almost anywhere in Weston as long as you water it well, and people do. But it is rarely economical to irrigate real pasture, so the key to productive grazing land is finding soil that holds water through July and August, when the weather can get very dry. In this part of the world, soils good for tillage are often not at all good for pasture—cropland tends to occupy light soils formed in glacial outwash of sand and gravel that drains too readily to keep grass growing through a dry summer. The grass on such soils starts out fast in the spring but can't carry much stock after the Fourth of July. Elsewhere we have pastures on very stiff, stony, unplowable slopes that don't look very promising, but miraculously, in spite of drought, the grass grows right through August. It turns out that New England's much-maligned glacial till uplands can provide excellent land for pasture.

Not all till was created equal. Some is shallow and studded with bedrock outcroppings and cannot grow much of anything except very patient trees. But other till is mantled deep over the bedrock and frequently has a nearly impervious hardpan at about two feet, permanently compacted millennia ago by the weight of mother ice. Water slides downhill above this pan, allowing these slopes to remain green and grassy long after neighboring soils have dried up and gone to sleep until autumn. A farmer wouldn't care to plow and cultivate such bony land every year, but (unless it is *too* wet and spongy) it is excellent for perennial crops like trees and grass.

A second good location for pasture, we found, was in swales that had enough drainage to keep them from being out-and-out wetlands. These spots tended to be too cold and wet for early spring grazing but became perfect by late summer, collecting enough seepage from the surrounding slopes to grow grass even in a severe drought. Many of these little fields were once the prized hay meadows of early farmers. The few that have not reverted entirely to swamps and marshes as the drainage system was neglected in the course of the past century can make excellent pastures.

A good part of every New England town is suitable for pasture. The difficulty remains what it always has been: maintaining a good pasture sward. There are few native grasses and legumes that are well adapted for this purpose in our region, but the European forage plants that have long been naturalized here do very well: bluegrass, timothy, and white clover among them. The key to the long-term productivity of a pasture is to encourage a vigorous perennial legume; white clover is the most familiar. It lies at the heart of good pasture because it fixes nitrogen from the air and fetches phosphorous and potassium from the mineral subsoil, making these nutrients available to the surrounding grasses through the constant sloughing off and decay of its roots and through the manure of animals that eat it. If a pasture contains a healthy stand of clover (or some other legume such as birdsfoot trefoil), then the old fancy that livestock fertilize the land as they graze it comes a bit closer to being true. In this case, the clover acts as a continuous pump to draw nutrients from both above and below the surface of the ground into circulation in the topsoil, replacing those that are removed as milk, wool, and meat. For clover to thrive, it is essential every few years to spread on each acre two or three tons of lime.

Granted a good stand of grass and clover and the faithful application of lime, improving the quality of a pasture depends on proper grazing management. Sod-forming plants thrive best when they are allowed to reach a height of a few inches, are grazed off quickly and evenly, and then are rested for a week or longer to grow and photosynthesize again, replenishing the roots—the process is very much like mowing a lawn. In fact, mowing a lawn simply imitates the movement of herds to which these grasses have long adapted themselves. Rotational grazing concentrates the livestock in a small area for a day or two so they graze it thoroughly and then moves them on to the next small area. As noted earlier, we used portable electric fencing to subdivide our pastures. Moving sheep, fence, and water every other day was work, but in most cases we were pleased with the way the grass responded. In pastures where the vegetation was poor, the ewes beat back the encroaching brush and stayed in tolerable condition. Our best pastures were partitioned into five or six sections, providing the lambs with fresh, tender grass every two or three days. Desirable forage species seemed to thrive under this regime and steadily expanded their territory year by year. Even on droughty soils, during wet summers the white clover magically spread into patches that had been grazed low. The challenge was to find the grazing regime that worked best on each soil. Improving pastures is a lifelong proposition, requiring decades of attending, pondering, and fiddling around.

As we moved our sheep around Weston, clearing brush, uncovering stone walls, and watching pasture come back to life, I got the feeling we were slowly learning an essential skill that had eluded New England farmers for centuries. For too long, pastures in the Northeast have been regarded as mere exercise yards for livestock fed mainly on alfalfa hay and corn silage, along with feeds from the Midwest. Because poorly managed pastures have so little feed value, most of them have been abandoned. During the past decade, however, under the new dispensation of rotational grazing, pastures have been making a quiet comeback. Letting the animals walk to their food is proving more economical for some progressive farmers than relying entirely on chemically grown, mechanically harvested feed. It will prove abundantly economical as the hidden costs of fossil fuel and soil erosion in conventional agriculture are more forthrightly accounted for in the future. That day cannot come too soon for the last struggling dairy farmers on the hillsides of New England. Given plenty of lime, good legumes, and careful grazing management, grassfed livestock can once again become an important part of our rural economy.

Even livestock that walk to their forage in summer need stored fodder brought to them in winter, which traditionally has meant putting up hay. Today, most local hay is grown on permanent hayfields that are cut twice and occasionally three times a year. The best of these are regularly limed and fertilized and periodically resown with red clover to maintain a proper mix of legumes and grass. We bought most of our hay in the field from our community and educational farm comrades at Drumlin Farm and Codman Farm in Lincoln, giving us plenty of grunt-level experience bucking hay bales into trucks and wagons and then up into the barn loft back at the Earles'. We loved putting up hay. The hard push to get the hay baled and off the field while the weather holds is always one of the high points of the farm year. Packing bales away into the sweltering upper reaches of the barn is an annual rite of passage in farm country, and the young and strong revel in it. In late summer, as the days grow shorter and the dew lies longer in the morning, it is a great satisfaction to get in a last cutting, to smell the loft safely filled with green, leafy second-cut hay, and to know that the ewes will be eating well when their lambs arrive in February and March.

In my mind, hay lies at the heart of mixed husbandry, where grassland and tillage meet. It is the key to their integration and to the flow of nutrients within a diversified farm. Hay can be grown both in conjunction with pasture and in rotation with tilled crops. In New England, there is often overlap between pastures and hayfields because the amount of grass that is available changes through the growing season. Grass grows fast in May and June but slows dramatically in the hot, dry days of July and August. Carrying a given number of stock through the spring requires only half as much pasture as will be needed in the late summer months. The obvious solution to the resulting spring surplus is to mow about half of that pasture for hay in June. In fact, this was the normal practice in New England until about a hundred years ago—before the advent of mechanical mowing it was unusual to take more than a single cutting of hay in a season. The second growth, or rowan, as it was often called, from these fields was then available for late-season grazing. The reliable aftermath of low-lying meadows became especially important in a dry year.

Making sure livestock have adequate hay, however, is not quite as simple as taking one cut from half of one's pastureland. In the first place, this alone would not produce enough hay to carry the pastured stock through the winter. In the second place, this would not produce good enough hay to satisfy certain animals. First-cut hay is coarse, grassy stuff, generally long in the stem and short in leaf and legume content. It is a suitable foundation for a winter ration for beef cattle and workhorses, perhaps, but is too low in protein for cows in milk or ewes in lamb. Our sheep throve better on second-cut hay, rich in clover or alfalfa. To produce such high quality legume hay, some land must be devoted to hay production right through the season.

In the best agricultural tradition across North America, legume hay was grown in rotation with grain crops. A typical rotation began with grassland plowed for a year or two of corn, heavily manured, followed with a few years of small grains such as wheat and oats, and finally returned to alfalfa or clover for several years of haying. Such rotations had the virtue of regularly rebuilding soil nutrients and organic matter. They kept the manure on the farm because livestock were included in the scheme of things. Even on cash grain farms with few cows, hay rotations were strongly encouraged by the need to grow fuel for the workhorses. When crops are being exported, it is difficult to maintain soil fertility by internal legume rotations and manure alone, but such tight nutrient cycling does go a long way toward putting a farm on a sustainable footing.

Unfortunately, fewer and fewer farmers practice this kind of beneficial mixed husbandry any longer. Grain producers focus instead on the production of one or two cash crops, heavily fertilized, which they have the expensive specialized equipment to cultivate and harvest on a large scale. Livestock producers, including modern dairy farmers, sit at the receiving end of the grain pipeline. They often concentrate so much purchased feed at their facilities that manure disposal overwhelms the capacity of nearby fields and becomes a pollutant—a bizarre perversion indeed. At that point, one is reluctant to call such places farms—they have truly become factories. One day corporate ownership completes the transformation, and then even the gloss of family farming is gone. Only the Amish, as a group, adhere to traditional rotations, keep a balance between livestock and crops, and continue to thrive as small diversified farmers. But when the full cost of industrialized farming is realized and the price of conducting agribusiness rises, such virtuous necessities as crop rotations, on-farm manure recycling, and mixed husbandry will be reimposed upon modern agriculture. This will eliminate much of the competitive disadvantage that has plagued small farmers and those farming on so-called marginal soils for more than a century. New Englanders will have to learn to farm all over again.

At Land's Sake we took some steps in this direction but have left a lot of ground to cover. This became clear to me as I thought about the complexities of integrating a grass and livestock program with our market garden, constructing a larger outer circle of land use around an inner one. We have long incorporated winter cover crops and leguminous green manures into our fruit and vegetable rotations to improve soil quality. The next steps would be to add livestock to the system: utilize some of those winter cover crops as season-extending forages and extend some of the green manures into full-fledged hay rotations. Timothy and clover could be rotated very nicely with our field crops, pumpkins, winter squash, potatoes, sweet corn. If a small grain such as oats were introduced into this rotation, we could grow some of our own feed and straw bedding. The final steps would be to cultivate the farm primarily with horses and, to close the circle, carefully apply manure from the barn at key points in the rotation.

Here is the outline of a workable agroecology for New England: grass, legumes, and livestock suited to the climate and to many of the region's hard-to-till upland soils would surround and support more intensive vegetable production on better tillage soils. Mixed husbandry and market gardening would overlap and mesh, like a large wheel that turns slowly and drives a smaller spinning wheel within. This is the direction in which New England farming was evolving a century ago but was brought up short by poor grass management on the one hand and a flood of cheap grain from the Midwest on the other. The motive force of sustainable mixed husbandry must be strong legumes within both pastures and hayfields, which directly replenish soil nutrients and increase the supply of manure available to fertilize grains and marketed produce. To really flourish, the grassland will require regular liming and probably periodic soil amendments of phosphate and potassium from outside sources. It is unlikely that New England will be able to produce all of its own feed grains in this way, so purchased feed will continue to augment locally grown hay and grain. This will bring extra nutrients to farms. It will be up to midwestern farmers to produce that grain sustainably, and up to New England farmers to pay the price that reflects that care. The system can be completed by recycling of composted municipal wastes to local market gardens and hayfields, giving the wheel another spin.

Land's Sake has not put together a model of such a complete system. Instead, after nearly a decade of keeping sheep and a few years of experimenting with draft horses, we made the difficult decision of putting our livestock program on hold. One reason was that running too many experimental projects at once was expensive, and we needed to concentrate on our financial strengths. Another was that we could not afford to invest in haying equipment. Above all, we lacked a barn to stand at the center of such a grass and livestock operation. Mixed husbandry in New England may have potential ecological virtues, but at present it remains economically marginal.

ECONOMY

Faith and I increased the size of our flock until we had thirty-five breeding ewes, about all the Earles' barn cellar could comfortably hold. The ewes produced some fifty lambs every spring, and because we usually carried a dozen or so yearlings as possible replacements, with the ram and his companion (a low-birthweight miniature ram named PeeWee) we had about one hundred sheep to lead to pasture every summer. The operation kept us hopping among a dozen different places until September, when we consolidated the flock for breeding. After nearly a decade we felt we had some idea what we were doing. The lambs were growing to a nice size, and we had a ready market and a good price for the meat and the tanned skins through the Land's Sake farmstand. We were meeting our cash costs for hay, grain, and medicines, and Faith was piling up lovely fleece and yarn as a dividend. The trouble was, we were getting hardly anything except that wool and a few lamb chops for our labor—and that we could have had with far fewer sheep.

Our flock had grown too large to remain just a pleasant avocation, but too small to be a viable commercial operation. To make it a paying proposition, we needed closer to one hundred breeding ewes, triple the size of the present flock. At that size, if run very well, the sheep might bring in enough income to cover the salary of a half-time shepherd. But neither Faith nor I had any inclination to become an official half-time shepherd, which we knew would be a de facto full-time job. She was already a full-time teacher, and I was trying to run Land's Sake and complete my dissertation. The project seemed to make the most sense as a part-time responsibility of one of the Land's Sake farm staff, but Land's Sake had no barn, no hay project, no good way to integrate livestock into its overall program. So Faith and I purchased the sheep and fencing from Land's Sake, reduced the flock to a manageable dozen or so, and let all except a few of our best pastures go back to the brush hog.

Feeding the flock. Wary ewe at left is Chocolate. (Kippy Goldfarb/Carolle Photography)

The grand experiment at turning a wealthy suburb into a giant sheepwalk was over. The great Weston Clearances that I had secretly envisioned—in which all the CEOs and their huddled families would be evicted, bundled into their Land Rovers, and exiled to Tucson, leaving sheep to wander over the crumbling ruins of their tennis courts and patios—unfortunately never came to pass. I won't say that the experiment failed because it did give us a great deal of pleasure while it lasted. Keeping sheep in the suburbs is by no means a lost cause. Ellen Raja and her husband, Roy, in neighboring Lincoln, who taught us much about shepherding, have carried on an operation much like ours for nearly thirty years. Like us, they break even or turn a small profit. Keeping sheep does not support many farmers in New England—most shepherds we know have some other source of income. But it is a rich way of life that more than pays for itself.[13]

About the time the sheep project began winding down, Land's Sake embarked on another livestock experiment. For several years we tried draft horses, with a handsome team of Suffolks, Roman and Lady Di, supplementing our tractor. Suffolks are a sorrel breed that is a bit smaller and more economical to feed than Belgians or Percherons. Roman and Lady Di were brought together in the early nineties by the farm manager, Steve Miller, and the forester, John Potter, who aspired to be teamsters. We used them both singly and as a team to pull logs in the woods and to plow and cultivate on the farm. The horses did a particularly good job for us in planting, hilling, and digging potatoes. They also pulled a hay wagon full of riders on fall weekends at the farm.

I was impressed with what the horses could do: when they were working well, they were quite efficient. They fit our philosophy of increased local self-reliance perfectly. There seemed to be potential for steady income by giving hayrides and by breeding more Suffolks. But the horses had drawbacks, too. They were expensive to maintain, and their care took a lot of time, even though Steve and John voluntarily stretched their hours to do chores. It seemed quite difficult, under our circumstances, for one farm manager to both find steady work for the horses and attend to pressing farm tasks in a timely manner. Like many things we tried at Land's Sake, the Suffolks began to look like a good idea whose time had not yet come—we did not have a barn or a hay and pasture program. We had a small staff that was scrambling to stay on top of our many projects as it was, without the distraction of incorporating horsepower. When John and Steve left Land's Sake, we were faced with the dilemma of finding new managers who were also teamsters. Needless to say, experienced teamsters are rare. It seemed we had gotten the horses too far ahead of the rest of the program. Reluctantly, the Land's Sake board decided to sell the horses and their gear.

The Suffolks, like the sheep, were a good idea that fell short on the first try. Someday, the revival of mixed husbandry in New England may be crowned by the return of workhorses to the farm and to the woods. Horses have their place, but recovering the skills and assembling the implements to use them effectively will be a long, slow process. Horses will show their real superiority to tractors for many farm tasks when the price of fossil energy rises. Even now they are economical for certain jobs and pleasant to work with. I hope that in the future, when we have a barn and a bit of pasture, Land's Sake will incorporate horses into the program again. Eventually, community farming may be a perfect means for returning workhorses to New England farming and for training young adults in their use.

One thing Land's Sake has not tried is dairy farming, and it is no wonder. Today, dairy farming is big business, and there is apparently less and less room in it for the small or even medium-sized producer. Until recently, dairying was far and away the agricultural leader in Massachusetts, as throughout the Northeast. Commercial dairy farming has been the backbone of the rural economy in New England for nearly 150 years, and as it wanes, so does the last of the area's active farmland. Milk farming has been on the skids in New England since World War II and in freefall for the past decade or two. Facing national overproduction and low prices, farmers on more expensive land in Massachusetts have great difficulty competing with the larger, more efficient herds that have been assembled in the Midwest. In spite of strenuous efforts by the Massachusetts Department of Food and Agriculture to support prices and promote the local dairy industry, milk farmers in the state have gone steadily out of business.

I do not pretend to be an expert on dairy farming, and I do not know how these farmers can be encouraged to stay on the land in the current economic climate. The immediate future does not look bright. But as I have argued throughout this book, the current economic climate cannot prevail forever. Demand for dairy products will rise as population grows, and the cheap energy subsidy that makes large-scale, centralized dairying economically feasible will someday disappear. Mountains of feed grain, seas of animal waste, and epic voyages of perishable milk products will give way, let us hope, to more decentralized, sustainable production. The advantage will return to those who feed their animals primarily on well-managed forage, minimize their dependence on harvested feeds, and supply local and regional markets with fresh milk. Specialty products such as goat cheese and ice cream will expand the market. Already, producers of organic milk are gaining market share, and some dairy farmers who rely on rotational grazing are flourishing. In time, these small but encouraging trends may come of age and reverse the decline in the dairy industry.

Meanwhile, a successful hay and livestock operation is not out of the question for a suburban community farm. In the neighboring suburb of

Lincoln is Codman Farm, the older community farm from which we lifted Land's Sake's bylaws. Founded in 1973, Codman has never done much market gardening or forestry but has concentrated instead on hay and livestock. Like Land's Sake, Codman is a product of its circumstances: it revolves around a historic barn that fell into the town's hands and utilizes the many fields preserved in Lincoln. Codman cuts hay on about 150 acres of Lincoln open space, selling the bulk of it to local horseowners. But Codman also has cattle, sheep, and pigs, their herd featuring such so-called minor breeds as Lineback and Devon cattle. The operation suggests another important role for community farms: to serve as a repository for thrifty older breeds that may prove invaluable to small-scale New England farming in the future. With help from donations and an annual harvest fair, Codman has now kept their version of community farming afloat for a quarter century. Stockraising in the suburbs is less than an easy road to profit but more than a pipe dream.

EDUCATION

There is great value in exposing people, especially children, to animals. I am not talking about field trips to the farm or visits to the petting zoo—although on festival days at the farm and at the arts and crafts show on the town green we did put a few lambs in a pen, and Faith gave spinning demonstrations. Every spring Faith took a pair of bewildered lambs to spend the day at her preschool. We always tried to live up to our reputation as providers of rural character. All that was fine, but of limited value. What I am talking about is kids taking part in the day-to-day, nitty-gritty details of raising animals for food.

A few years after we got the sheep flock going, several young families moved into our corner of town. They were delighted to find something resembling a working farm in their neighborhood and would drop by the Earles' barn on weekend walks to visit the lambs or watch the shearing. One afternoon a pair of mothers appeared with a flock of their own offspring just as twin lambs were being born, and the kids witnessed the whole production from the emergence of the water bag to the ewe licking the little ones clean and getting them up and sucking. The children were fascinated, the mothers transported—they had just scored a major triumph in their suburban mothering careers. And I was asking myself the same question I find myself asking regularly: through what kind of blind stupidity have we eliminated this kind of experience from the daily lives of most people?

Children are emotionally able to handle the earthy details of keeping livestock. One time I was walking through the barnyard with several ten- and twelve-year-olds as we fed the flock, and they were mincing and snickering about stepping in poop. Suddenly I saw something unusual and a bit alarming. "Now look here!" I exclaimed, squatting down. A long, white tapeworm was shining in the droppings. Immediately the boys and girls snapped into a sharply inquisitive mode, squatting down beside me, examining the worm closely, looking around for others, asking analytical questions, not the least bit squeamish anymore. Were the sheep going to be all right? That, quite properly, was their main concern. I explained that probably one of them had spit up her pill the last time we wormed or had somehow been skipped—maybe I had checked off the wrong number on the list. The next worming was coming up soon. Meanwhile, Faith and I would be checking the flock to see if any sheep was listless or off her feed. The children began to do that at once for themselves. And I thought, now these kids aren't seeing just a bunch of cute, amusingly scatological sheep—they're focused on the *condition* of the sheep.

One spring a neighborhood girl took a special fancy to a bottle baby, named him Fleece, and helped feed him. We reminded her that most of the ewe lambs and all of the ram lambs were going to be slaughtered and eaten in the fall. She said she knew that, but that Fleece was cute anyway. And Fleece was cute: he had a long, handsome nose, a thick, colorful fleece, distinctive knock-knees; and, being a bottle lamb, he was very friendly. We didn't allow the girl to make a pet of him, but she did visit him from time to time during the spring and again when the lambs returned to the Earles' in the fall to be finished on the oat forage we had planted. Now it is a poignant peculiarity that as they mature ewe lambs grow shy and skittish, but ram lambs going through adolescence become both more aggressive and more affectionate. They come up and butt around playfully. Fleece was an extremely playful little chap.

Spinning yarn at the Case Farm: Faith with her wheel (Susan M. Campbell).

Faith looked through her records but couldn't find a single legitimate reason for keeping Fleece: he was a nice-looking, growthy fellow with good bloodlines but not outstanding enough to become a ram. Anyway, we didn't raise rams from our own flock for breeding—instead we bought a new ram every two or three years to keep the flock from becoming too inbred. (The rule is you can breed to daughters but not to granddaughters.) We had no living use for Fleece, and neither did anyone else. So Fleece went off with the rest of his cohort. I won't pretend taking the lambs to slaughter was ever a joyous occasion for Faith and me, but that November day was particularly somber. We wondered if we would ever see our young neighbor again. But next spring she was back at the barn with a couple of friends from school,

Spinning yarn at the Case Farm: Carol LaLiberte teaches a girl to spin with a drop spindle (Susan M. Campbell).

telling them matter-of-factly all about her friend Fleece who had a good life but ended up as lamb chops.

I do not endorse the conditions under which the vast majority of meat animals are raised in the United States. Yet I do not think that vegetarianism is the only legitimate response to feedlot confinement, any more than that wilderness is the answer to the poor way we raise most of our vegetables. It is good for children (and adults, for that matter) to learn that it is possible to treat animals with affection and respect and still kill and eat them. In my experience, participating in the slaughter of animals does not necessarily make one hard and brutal, but it can make one thoughtful and conscious of power and of consequences. I do not believe that it is morally necessary for everyone who eats meat to personally kill animals. I do believe that for people who eat meat there is great value in coming to know and care for animals, so that the act of eating is not completely divorced from the lives and deaths of the fellow creatures on whom we depend. It is a cliché these days to say that kids don't know that milk comes from a cow; or that if they do know it abstractly they have never seen it happen. I'm not sure if that is true, but I am quite sure that very few children know that meat comes from cute little fellows named Fleece, and I am confident that the ones who do know are better off than those whose concept of violence and death is limited to Rambo movies and Saturday morning cartoons, where killing is easy and fun.

A friend of ours who keeps livestock had no problem with her children learning firsthand where their food comes from, but she wondered about letting them witness slaughter when they were too young to understand what was going on. One afternoon she left her three-year-old daughter napping in the house and went outside to kill and clean a chicken for supper. When she was halfway through the job and up to her elbows in feathers and gore she heard a sound, looked around, and there was her child, watching the whole process with eyes wide open. "Oh, great—I wonder what she'll say now?" thought the mother.

"When do we eat?" asked the little girl.

A livestock program should be part of every community farm. Even without extensive pastures and a rural setting, it can be done. A few animals, their feed crops, and their manure can be effectively integrated into the soil-replenishing rotations of even small market gardens in the suburbs. This greatly enhances the educational value of the farm program by what it suggests about sustainable food production.

A few miles down the road from Land's Sake is another sibling organization, Natick Community Farm. This farm grows almost every crop and creature known to agronomy, on a mere three acres next door to an elementary school. Everything is run organically—in fact, Director Lynda Simkins and her crew have long been among the leaders of NOFA. At Natick Community Farm, children come in daily contact with chickens and eggs, sheep, cows, goats, and pigs. These animals are not pets—they produce and become food. Former Codirector Martin Gursky has put together a fine community farming handbook that describes how they do it in Natick. We are in awe of their ability to make so much happen in such a small space. Not every town may have the hundreds of acres of open space needed to run a full-scale farm operation, but there is no reason every place should not have a vital, buzzing community farm.

ESTHETICS

People need many different ways of reinforcing their bonds with the land to guarantee that their souls develop an ample capacity for affection and care. This is something that not only bears repetition; it requires repetition. Coming to know and use a place responsibly is connected to slowly perceiving in an ordinary landscape a beauty that is more than scenic. It is a great pleasure for me to wear clothing—homespun sweaters, hats, and mittens in rich browns, creams, and grays—made by my wife from our own sheep. Faith wears woven scarves and shawls, as do many of our friends. She dyes some of her yarn with native materials, so other subtle colors of the landscape shine through. This is no shaggy, lumpish stuff still tangled with alfalfa stems and smeared with manure, but warm, water-repellent apparel that is both handsome and comfortable. I like glimpsing in the colors and patterns of these garments a reminder of our multicolored sheep, dispersed on their summer pastures or surrounding a hay feeder in the snow. It is like having an aspect of the landscape on my very person.[14]

Our house is littered with sheepskins, as are those of many of our friends. Besides being nice to look at, sheepskins are soft but firm and springy, nice to sit on and to lie on. Physical therapists swear by them. I can guarantee their efficacy for a bad back, especially one caused by lifting a big lamb into a pen improperly. We have sheepskins on chairs, sheepskins on car seats, sheepskins for exercise mats, second-rate sheepskins for bedside rugs and dog beds. Some of these pelts come from sheep we remember fondly. All of them keep us in mind of the way we live outside as well as in. Chocolate, the champion high hurdler I mentioned earlier, is retired now: I am sitting on her.

Best of all is the lamb itself, the meat. Americans are not great lamb eaters, which is a shame because there is nothing like the taste of grass-fed lamb. I knew that locally grown vegetables usually taste better than store-bought because they are picked in their prime and eaten fresh. But how would that apply to meat, especially since all of ours was flash frozen? Years ago I had read in an essay by the Iowa writer Vance Bourjaily that the meat generally available in the marketplace bears little resemblance to what some knowing farmers raise for themselves; but I had no way of appreciating the truth of this claim until I tasted our lamb.[15] What a revelation! No special preparation was needed—a simple roast with a little garlic and rosemary was startlingly flavorful. What had I been eating all these years?

I'm not sure exactly why our lamb tasted so good. I do not believe it was just self-delusion brought on by raising it ourselves, although that may have had something to do with it. (If that is what it takes to make food taste immeasurably better, then it is well worth the trouble, and I heartily recommend it.) It may be because our lamb was raised almost entirely on grass rather than grain, making it not only flavorful but lean and healthy, too. Chefs tell me this gives lamb a very clean taste. It may be because we slaughtered our lambs a bit younger and smaller than the blocky behemoths that dominate the market. Or perhaps the mix of wool breeds that made up our flock actually had sweeter flesh than the big meat breeds. Whatever the reasons, we learned once again that the vaunted modern food system is not delivering the best meat to our local grocery any more than it is providing the best vegetables. We don't buy lamb in the supermarket anymore, even now that we are not raising our own. It doesn't taste like much of anything, no matter how well prepared. Though not quite as bland (or faintly fishy) as what passes for chicken in the stores these days, it isn't really worth eating. We buy our lamb from friends who still raise sheep.

Faith and I did not get into farming on a back-to-the-land, self-sufficiency trip. We do not industriously can, dry, and freeze our produce for weeks on end so that we can completely wean ourselves from the supermarket habit. We do not raise and kill all our own meat. Not that I am opposed to people living in this honorable way if they want to and can organize their lives to have time for it; I'm just repeating that the purpose of this book is not to urge everybody to grow all their own food. We did eat quite a lot that we grew ourselves, and I have to admit that this was a major benefit of being involved with Land's Sake. As a born-and-bred skeptic, I found myself being constantly amazed at how much better the things we grew tasted than the food I had been accustomed to. It began with the carrots and cider I polished off the first day I worked at Green Power, and happily that did not turn out to be a figment of my imagination. Sometimes, without planning it, we find ourselves eating meals made almost entirely of homegrown stuff, and I discover not just my plate but the whole table, the whole room, the whole world glowing with flavor and vitality.

These meals are not fancy, but they often have a synergy that makes them more than just delicious. They are a celebration of the diverse elements of our lives. Picture a Saturday evening in October, after a day spent delivering firewood, pressing cider at the farm, or bringing the flock home to the barn—ah, the suburbs! The evening is cool enough for a fire in the woodstove, and old friends who have come to the farm to help out have stayed for dinner. The roast lamb, from our own pastures, is superb. Along with the lamb go roast potatoes from the little field at Green Power where we cleared the abandoned orchard in 1977 and where I gave Faith a book of Robert Frost poems. Not any old potatoes but a rich, flavorful variety like Yukon Gold—or maybe some nutty little Jerusalem artichokes in their place. Every year it seems that one variety of winter squash is sweeter than all the rest. The handsome Sweet Dumpling delicatas, the clear winners this fall, don't need maple syrup—they're sweet enough as is. For salad, a mix of fall greens—baby lettuce, arugula, mustard, and mezuna, graced with the last of the spicy red and yellow nasturtium flowers. Nothing complements Weston lamb better than dry blueberry wine from nearby Nashoba Valley Winery, aged in oak; or perhaps fresh apple cider, a tart pressing into which we have mixed a few bushels of rare Roxbury Russetts. For dessert, vanilla ice cream topped with Heritage fall raspberries, small but sweet.

A meal like this suggests a complex relationship with the land, one that is more satisfying than gardening alone can encompass. Many people who move to the country plant a garden, and that is good. Gardening is a fine start in restoring connections with the land and the seasons, but why stop there? The garden occupies only a small part of the landscape, engages only the growing half of the year, and provides only a fraction of our sustenance. To reap a full sense of place in New England a person does better to be involved in three activities: growing a garden, keeping livestock, and cutting firewood. Certainly other pursuits could be added, but these three nicely cover the ground. The garden affords the intimate experience of manipulating nature intensively, yielding fruits, flowers, and vegetables during the warm months. Keeping livestock introduces longer rhythms that carry right around the year, in summer cutting hay and managing pastures, in winter feeding hay and attending the births that usher in another season. Husbandry adds a closeness with animals to the garden's cultivation of plants, meat to the vegetables at table, and pastures and hayfields to the view. Cutting firewood is primarily a winter pursuit, working one cold season to supply heat for the next. It occupies the wildest, least cultivated part of the landscape and puts us in touch with the forest, our most important repository of natural diversity and guardian of ecological health. Gardening alone cannot capture the complexity of coming to know the land in this way, neither literally nor literarily. I believe involving oneself fully and responsibly with the life of a place is beyond any simple metaphor. To do so is to weave a complex, elastic web of many strong intellectual and sensual strands that bind us to the place where we live, whichever way we turn.

Again, I do not advocate that everyone grow a large garden, keep a menagerie of creatures, and tend a small woodlot on their own five- or ten-acre suburban lot. If people want to do that, I wish them well. More to the point, more properly scaled would be building *communities* in which people live in the midst of such activity and sometimes take part in it. As people with pastoral yearnings but more practical earnings resettle rural areas, as they are now doing in droves, they need to actively support those who still do make their living from the land. They need to see to it that working farms and forests still cover the major part of the landscape. They need to protect what they reinhabit from themselves. Then people who do not farm will at least live among people who do, appreciating the landscape and purchasing local products from their neighbors. Young people growing up in these places can find valuable work helping out on the farms. But such communities will flourish only if we stop insisting that farming is solely an economic activity that must be done in the most efficient way possible in order to produce the cheapest possible food and begin valuing well-cared-for, attractive land, healthful natural products, and healthy work as broader cultural benefits. We can promote these ecological, educational, and esthetic values of farming even further by organizing a community farm in each place to be a means for those who are not professional farmers but want to be involved in farming, and especially for children, to develop the subtle connections to the land that are required to take good care of it. Such engagement could be the basis for a new agrarianism.

Right now, most people who move to rural places are damaging them. Their very arrival is often the death knell of the rural sense of community and attractive landscape they seek. We cannot garden our way out of this dilemma. The home garden alone is too much an inward retreat to the security of a small piece of property where one makes a private peace with nature. But that peace is an illusion. Gardens are perfectly compatible with suburban sprawl, and therein lies their limitation. Those who move to the country because they love the countryside have a responsibility to help bring back not just gardening, but farming. We need to turn and meet the advancing suburbs head on. The suburbs must be transformed into something better: new kinds of agrarian communities, places that enjoy common agricultural engagement. We should not be talking any longer of garden suburbs but of the rebirth of agrarianism, of complex, pastoral communities.

Is such a revival of mixed husbandry in Massachusetts possible? Is there any room for it, and would it be ecologically and economically sound? On the one hand, it is good that New England's forest has recovered in the past century, and I am thankful for it. Far too much land was cleared for farming in the early nineteenth century, and some of it was too thin and bony to be sensibly devoted to growing anything but trees. There is no question that we need to keep a substantial portion of our landscape in forest. On the other hand, not all of the farmland that was cleared and subsequently abandoned in New England was inherently unsuitable for agriculture. Little of it may have been great tillage land, but much of it could have been excellent grassland. Potentially productive pastures were abandoned first because they were mismanaged and hence temporarily exhausted, like a runner in a marathon who foolishly starts out at a sprint; and second because in this condition they could not compete at feeding livestock with the cheap grain that, thanks to extractive industrial agriculture, was pouring in from other regions.

If a significant part of this land could be sustainably farmed, then we need to ask seriously how much of it should be growing food. Who says the proper ecological state for *all* of New England is forest, any more than that the Midwest should all be put back to prairie or the irrigated West all returned to desert and steppe? Those who celebrate the return of the northeastern forest as a parable of ecological recovery are remembering only half the story. They are ignoring the withdrawal of productive engagement with the land that made recovery possible. As a consequence, they are overlooking the near eradication of the tallgrass prairie ecosystem, which, in effect, allowed the eastern forest to recover. We have to grow food somewhere—in fact, we will soon need to grow more food everywhere. We need to determine a desirable balance between land in natural ecosystems and land in sustainable agroecological systems in every region. In New England, a substantial part of the landscape, now agriculturally idle, could again play an important role in a diverse system of pasture and livestock farming. To say that it is not farmable because it is not suitable for wheat combines displays little more than a misplaced confidence in the ability of wheat combines alone to feed the world.

How much of this land could be farmed without cutting back the forests too severely? An impossible question to answer precisely—but we can at least quickly confirm that there is still room to move in this direction, even in a small, heavily populated state such as Massachusetts. Currently only about 235,000 acres, or a mere 5 percent, of Massachusetts is counted as improved farmland. About 65 percent of the state is forested, which means the remaining 30 percent of the land is developed. Conservation biologists argue that for safety's sake one-half of the landscape should remain in permanently protected forest: perhaps more forest than we strictly need, but let us proceed conservatively. We can see in a very crude way that the amount of active farmland in the state could be quadrupled from 5 percent to 20 percent, that is, to nearly 1 million acres, and still half the land would be in forest. Were we to devote 80,000 acres to market gardens, most of that million acres of active farmland would be left for pastures, hayfields, and other cropland. An increase of this dimension would be a dramatic change indeed, returning us to a landscape as open and farmed as during the early decades of this century. Here is ample scope for the revival of a healthy mixed husbandry system, a world of grass-based dairying, sheep farming and workhorses surrounding and supporting a smaller amount of intensive market gardening in town after town—especially those just beginning to suburbanize. This is the world to which those who love rural character should be laboring to give birth.

At present there is nothing for so much reborn farmland "economically" to produce. I am not suggesting that, having punched our fantasies into our calculators, we should step outside, fire up our chainsaws, and start making stumpland again. Massachusetts farmers won't be seriously competitive in producing milk, meat, and grain until the present parameters of food and land markets change. Four things will have to happen to make ecologically sound mixed husbandry in Massachusetts economically feasible. First, there will have to be a sustained rise in the price of grain, reflecting increased global demand and the mounting costs of industrial agriculture, particularly energy costs. Second, Massachusetts farmers will have to become superb pasture managers, coaxing sustained productivity from their grass. Third, a strong market for distinctive locally grown farm products like goat cheese and grass-fed lamb will have to be built. Finally, nonfarming rural residents will have to pony up to help protect more than token proportions of farmland—partly for what it can produce, but mostly for the well-being of the community and the landscape.

A national agricultural economy embracing these values would be every bit as legitimate, and a good deal more attractive, than the present one. The shift toward stronger regional food systems will not come overnight—it will be partly driven by global necessity and partly drawn by local desire. The resulting food may be somewhat more expensive, but it will be better—and a more habitable world will be created in the process. We can start making some of the requisite changes now, while we wait upon others that are beyond local power. Most important, we must change our attitudes toward residential land ownership and settle rural towns in ways that protect the great bulk of the land, be it farmland or forest. Given that, we and our children can gradually build a new agricultural economy until half of the land is back in farms and the forest tells us to stop.

A few years after we fenced his pasture, Dr. Earle decided to clear an acre of scrub woods that had been growing up in the back corner for decades. So, on a drizzly spring morning a crew of Land's Sake stalwarts moved into the dense trees and brush. Three of us kept chainsaws constantly running, while others hauled slash to the roaring bonfires we periodically ignited as we moved down the field. Cutting and burning brush in the rain is grimy, exhausting work but immensely satisfying. It is exhilarating to drag cumbersome branches and thorns a short distance and cast them into the fire, watching them be instantly consumed. The flames of a good brushfire twist upward orange and black, stirring deep ancestral memories. At lunchtime we lounged by the fire, chewing our bread and cheese and washing it down with cider. Then, saws filed and refueled, we were back at it. The weather cleared in the afternoon and we grew hot and weary but drove on to the end of the field. Pile on more and more, feed the insistent flames, keep it rolling! This is not the kind of job you want to leave unfinished if you can help it. We spared a few shapely trees here and there for pasture shade, but by evening the acre was clear. We sat by the last fire drinking beer and rolling the final pieces of blackened wood into the dying flames. When nothing but charred coals remained, we doused the fire and went home satisfied.

Over the next few months, as I had time, I bucked and split the ten cords of red maple and black cherry trunks we had left strewn over the ground and trucked the wood up to our house to supply winter fuel. I then disked the ground every month or so, bobbing and weaving among the stumps. In the late summer we sowed pasture seed, under a nurse crop of oats that fattened the lambs in November. By the following summer, the ewes were grazing a fine sward of timothy, bluegrass, and white clover. Years later, the sweet places where the fires burned are still marked by lush circles of legumes. At last, the entire pasture at the Earles' that we had enclosed almost a decade earlier was neatly cleared back to the stone walls, where the silver wires now ran. The grazing sheep beyond the antique house and barn formed a small gem from an earlier time, set among the encroaching forest and suburbs.

I suppose that some in the environmental community might regard with horror this tale of slashing and burning forest to make way for foul-mouthed sheep. After all, sheep have been the bête noire of the conservation movement since the writings of George Perkins Marsh and John Muir, the slayers of forests from the Cedars of Lebanon to the Sierra Nevada. Don't we need more trees in the world, not fewer? Yes, we do, in many regions and in total. But I am not talking here about the wholesale destruction of the tropical rainforest. New England is well forested, and I wouldn't have it any other way. I am talking about the judicious restoration of farmland *within* that forest. There is room to fit a few pastures and hayfields into an ecologically healthy landscape. By so doing and by increasing the amount of food we grow for ourselves within the region, we will reduce our dependence on the global industrial economy and thereby the pressure we place on other ecosystems around the world today.

Clearing land for pasture can be an act of witless environmental destruction or an act of willful environmental embrace, depending on the circumstances. The traditional New England landscape included fields and forest, and although we should take heed of that tradition's ecological flaws, such a landscape evokes a powerful emotional response in the people of the region. That instinct is good and needs to be nurtured. It is not merely nostalgic, if it can be acted upon responsibly. In the future, with perseverance and care, New Englanders can restore ecologically sound mixed husbandry to a significant portion of the Northeast, the better to keep the forest.

Weston Middle School student Maryann Belason drives a tap into a sugar maple. (Thomas Gumbart)

Chapter 4

Tree Crops

This chapter is about more than tree crops. It is about all those parts of the agricultural landscape that lead people back toward the woods: apple orchards, sugar bushes, cranberry bogs, and blueberry swamps. Even though these New England icons aren't all strictly native or strictly woody, they are more closely related to the native forest than corn and tomatoes, sheep and hay. These native or strongly naturalized perennial crops occupy the creative fringe (both geographic and conceptual) where cultivation meets the wilder parts of the landscape, the forests and wetlands. Although such crops may cover only the outskirts of the cultivated ground, because they spring from the land's native bent they often give a regional culture its characteristic shape.

The cultivation of such crops could be considered agroforestry, although that might be stretching a term, too. In 1929, the Columbia University geography professor J. Russell Smith suggested in his classic *Tree Crops* that agricultural systems should closely mimic native ecosystems—in the case of

New England, the forest. This idea of "natural systems agriculture" lives on in the concept of agroforestry for the woodland region and in the efforts of the Land Institute in Kansas to develop polycultures of perennial grains for the prairie region. Harvesting more food from trees is certainly an idea worth exploring, even as we improve the practice of mixed grass and tillage husbandry. We would also do well to look closely at the ways of Native Americans, who lived here successfully for thousands of years before whites, employing foraging and horticultural systems that were more closely tied to the forest than the European husbandry that replaced them.

Tree crops and native perennials have a venerable place in Yankee agriculture. Syrup, cider, cranberries, and blueberries remain our most distinctive regional specialties and deserve to be celebrated. Whether or not we someday shift our entire agricultural system to a version of forest farming, we can encourage many more perennial species to flourish at the edge of the cultivated land, contributing ecological and culinary relish to this part of the world. In time, tree crops may be able to provide more than mere relish and move into the meat of our diet in this forested region.

Maple Syrup

The sugar maple (*Acer saccharum)* is probably the most valuable native food tree in North America, at least since the demise of the American chestnut. Its only rivals might be the pecan and the black walnut. The romance of sugaring is so strong that those who fantasize about going back to the land in the north country invariably sweeten their dreams with maple syrup. Sugaring is a pleasant and rewarding way to get out and enjoy late winter and that most cherished New England season, mud time.

Syruping in Weston begins in the middle of February and runs to the end of March. There is a fine art to judging when the time is right to hang the buckets: too early and the holes may heal over before the sap starts running, too late and the early flow may be missed and warm weather soon spoils the run. No year is ever like the previous one. Spring in Massachusetts is an on-again and off-again affair, and it isn't easy to guess whether February or March will bring the best sugaring weather. To keep life simple at Land's Sake we ignore all the natural signs and simply tap out at the beginning of

February school vacation, whenever it may fall, because our program is run mainly for the benefit of children. I had to compromise on this point early in my budding suburban syruping career.

The sap run is highly dependent on the weather and comes in several short bursts spread out over the six-week season. Sap flows when the days rise well above freezing and the nights sink well below—if the days turn windy and cold or the nights turn warm, the sap run ceases until conditions improve. Weston is close to the coast, so the weather usually warms up too fast. As the buds on the maples begin to swell, both the quantity and quality of the sap declines rapidly—it goes from clear and sweet to cloudy and then "buddy" (yellow and bitter), and that ends the season. Our part of New England is not normally cold enough to make large amounts of premium syrup, but in years with a late, cold spring we do well enough.

When Bill McElwain built the sugarhouse beside the new middle school in 1973, Weston abounded with large, seemingly healthy sugar maples. These ancient trees stood in solid rows along many roads and also lined the driveways of old houses and shaded their yards. It was a simple matter to drive a crew around town emptying the sap buckets into five-gallon plastic pails, and from there into a collecting tank that rode in a trailer behind Bill's jeep. In the early years of the project we would hang more than a thousand buckets every spring. The appearance of the buckets on the trees became an eagerly awaited harbinger of spring in Weston, as if this were something that had been happening time out of mind. The sap buckets made Weston feel like the traditional New England town that people liked to think they inhabited, a survival of something deep. The truth is, as far as I can discover maple syrup was never made in any quantity in Weston before Green Power.

For years, I pushed to increase the number of buckets that Green Power was hanging, partly because I was young and ambitious, but mostly because we possessed a big evaporator with a mighty thirst. The five-by-fourteen-foot twin-pan Leader evaporator was too large for our operation, requiring a minimum of three thousand buckets to keep it supplied. Once the sap leaves the tree, it must be boiled quickly or it will spoil. But a woodburning evaporator is not fired up with the flick of a switch. It took an hour or two to get a rolling boil across both pans and another hour or two to shut down at the end of the day; the evaporator could easily go through a couple of hundred gallons of sap per hour, which meant we needed to have a good thousand gallons of sap ahead to justify boiling. Collecting that much took a few days, so we stored the sap in a series of tanks behind the sugarhouse. An ideal medium for bacteria and yeast, the sap quickly fermented in the balmy flatland spring. We needed to hang more buckets so we could boil more frequently with fresher sap and thus make fancier syrup. At least that was my philosophy.

During the late 1970s we managed to tap more maples in Weston and neighboring towns. We peaked in 1980 at about fifteen hundred buckets. That spring marked a tidal change in the way the world was moving, in my mind—away from one kind of "green power" and back toward another, money. I vividly remember driving up through New Hampshire to pick up the last few hundred used buckets at a sugarhouse in Vermont: Ronald Reagan signs were everywhere and suddenly I realized that this cheerleader of mindless growth was going to win the primary and be elected president. I had a chilling premonition that the anticipated Age of Ecology was not yet upon us, maple syrup in the suburbs notwithstanding. We have not had to buy any more buckets in Weston since then.

At about the same time, we began to notice that one by one our maples were dying. It didn't seem to matter how old or young the trees were or whether we were tapping them or not. First, the leaves on some of the uppermost branches turned red and gold far too early in the fall, in August instead of October—a sure sign of stress. Over the next few years the entire crown typically began to look ragged and thin. Next, major sections of the tree died off and were pruned away by the tree surgeons. Finally, one day, the whole tree was gone.

During the 1980s sugar maples all over town began disappearing, like the elms half a century before them. By the end of the decade it was as if the maples had never existed—only a handful remained. Imported disease had killed the elms, but American road salt killed the maples. They had graced the small rural town of farms and estates that Weston once was but could not survive the growth of suburbia. Bill McElwain happened along just in time: had he shown up a decade later, the maple project could hardly have been started; had he appeared a few decades earlier, there would not have been enough mature maples to tap. The era of sugar maples along Weston's roads was startlingly brief.

Sugar maples are not native to Weston—or at best have been rare over the past several thousand years. They were uncommon trees in the native forest that greeted the English in the part of Massachusetts stretching from Middlesex County southeast to Cape Cod. The same was true of other species of the "northern hardwood" forest: beech, yellow birch, and hemlock,which are found only in small patches in our woods today, usually on steep slopes or in small ravines. There is no evidence that they were once common in these parts, only to be extirpated by colonial logging and clearing. Red maple (*Acer rubrum*) is much more abundant than sugar maple, but this species is inferior for making syrup. There is no evidence either that the horticultural Indians of southeastern Massachusetts made maple sugar. It was the foraging tribes who lived in the hill country to the north and west who showed the secret to the Europeans. The implication is that there were no suitable maples in this corner of the world for people to tap, going back not just for centuries but probably for millennia. Fossil pollen cores from Cape Cod do record brief incursions of northern species during the past ten thousand years, but oaks and pines have always reasserted themselves to dominate the landscape. The persistence of oak forest and the exclusion of sugar maple and its northern friends were probably caused by a warm,dry climate and enforced by recurring fires. I will return at the end of this chapter to the controversial use of fire by Native Americans to manage their environment and enhance their natural food sources.[1]

Whether because of prevailing fire or climate, sugar maple was conspicuously absent from the forests of colonial towns in eastern Massachusetts. We have maples in Weston today because our great-grandparents began planting them a century ago. Nineteenth-century photographs leave no doubt that elm was once the preferred shade tree for gracing Weston roadsides and dooryards, although sugar maples appear here and there. A new era of tree planting got under way toward the end of the past century, as part of the same beautification movement that led to the creation of the town common.

During this period local elites throughout New England were sprucing up their adopted rural communities. In this civic spirit, the Village Improvement Association of Weston was organized in 1895 by the town's leading citizens to "preserve the natural beauties of the town and improve and ornament the streets and public grounds by planting and cultivating trees." Town residents could become association members either by paying one dollar per year or by planting one tree. It appears that most of the town's roadside maples as well as many trees on private estates were planted in the course of the following decade or two. It is tragic and telling that these trees did not outlast the living memory of those who planted them.[2]

Why did sugar maples become the style of the twentieth century, in place of nineteenth-century elms? Maples may have emerged by default as elms began dying from disease in the early decades of the new century. But the preference for maples may have deeper roots. Sugar maples were probably chosen to improve Weston's streets for the same reason that a common was created for Weston's village center: they had come to be strongly identified with the classical New England village, even in such towns as Weston, where they were not actually indigenous. Civic leaders were determined to give their town the proper New England flavor, and that meant maple. In any case, maples were a happy choice and certainly did beautify the town. They were planted everywhere. In 1914, Tree Warden Edward Ripley reported that he had begun "a policy of setting shade trees, putting out the past year some forty rock maples," mostly along the Post Road. Many of the stately maples Green Power began tapping in the 1970s, which we regarded as immemorial, were in reality no more than sixty years old and had no local progenitor other than fashion.

In the era when sugar maples were being planted, most travel in Weston was by horse. The snow was rolled on the roads in the winter to smooth the way for pungs and sleighs. But as the maples grew up to replace the elms, changes that were to prove fatal to the trees came to Weston's streets, demonstrating the real priorities of our civilization. Telephone and electric wires appeared around the turn of the century—these eyesores were welcomed into town at exactly the same moment that the streets were being beautified by trees. It is hard to imagine two more contradictory impulses. As the shapely maples matured, they were disfigured to keep them from growing into the multiplying wires. Even worse was the increasing flow of automobiles, which became a flood after World War II. Suburban commuters demanded clean pavement, and so the snow roller was retired, and the highway department switched to plows, sand, and salt. By the 1970s the maple trees were dying as an unintended consequence of snow removal. Unfortunately, salt has proven to be highly toxic to the roots of many trees, maples in particular. The haze of automotive air pollution that shrouds the Northeast hasn't helped the trees, either. Today, roadside maples in Weston have all but vanished. One generation planted Weston's maples; the next generation promptly killed them off.

The maples were a casualty of the automotive suburb, and their passing demonstrates once again how the quest to commute from urban blight to rural sanctuary is self-defeating. The simple technical fix for the wires would be to bury them underground, slightly raising the cost of electrical service. But this would not protect the trees from road salt and air pollution. There may be technical fixes for these problems as well, but they grow progressively more expensive and thus demand a greater change in values. A biodegradable derivative of corn could be used to melt snow on the roads, but this is costly even now and ultimately depends on a surplus of grain produced by the unsustainable squandering of fossil fuel on our farmlands, and so gets us nowhere. Ultimately, we may find that to enjoy so simple a pleasure as roadside maples demands a much more dramatic change in the way we interact with the rest of nature than a series of simple technical improvements alone.

In early February 1978, a major blizzard swept up the East Coast just as Green Power was preparing to hang buckets on the trees. Boston and the surrounding suburbs were shut down for a week by nearly three feet of snow, driven by a fierce northeast wind. As soon as we could make it to the sugarhouse, we shoveled paths to the doors and dug out the woodpiles, sap tanks, and bucket shed. We hung the buckets the next week as planned, clambering up snowbanks plowed high alongside the trees, drilling the holes off our boot tops instead of our belt buckles to keep them low. In spite of these precautions, late in the season after most of the snow had melted, little kids had to reach high above their heads to lift full sap buckets from the trees. It was a comical and endearing sight and a memorable season.

In the days immediately following the storm people in Weston had a wonderful time. They pulled together to dig one another out, took care of shut-in neighbors, and displayed all the unremarkable generosity that naturally shows itself in emergencies. There were a few glorious blue-sky days when no one could get to work, to school, or to the malls. The haze of modern life had been swept away, and the worst of its commercial clutter lay hidden under pristine white powder. It was a true holiday because it was free of the frenetic travel that goes with modern family gatherings. The roads remained covered with snow, so there was little traffic to contend with. People came out of their houses on skis, put the kids on sleds, and filled the quiet streets of their neighborhoods, visiting one another. The town forest was thronged with enthralled skiers. "It was so nice to see everybody," the suburbanites all said. "We met our neighbors for the first time. It was so quiet and peaceful—such a *community*." Needless to say, the next week it was back to business as usual, as people raced to catch up at getting and spending, turning the world immediately outside their front doors back into a noisy, noxious, dangerous place.

Does it take a blizzard to bring us fleetingly to our senses? If we really value peace and quiet, visiting our neighbors more than once a year, and reveling in deep, clean snow, we could easily have them. We do not need to enjoy such things only once or twice in a lifetime and go through the rest of our earthly existence feeling wistful but staunchly realistic about our harried lives. It wouldn't particularly inconvenience drivers to plow only the state highways down to the bare pavement, to limit the speed on the main town roads to twenty-five miles per hour and the side streets to fifteen or twenty, year-round. Limiting our efficiency by slowing down would be as sensible as rushing to generate extra income so we can buy property up-country or getaway weekends at country inns to achieve the same end. This way, people could plant sugar maples along the roads again and not have to watch them die.

I offer these suggestions knowing full well that hardheaded realists will regard them as absurd. They are not as easy to realize as I make them sound because they imply a radical change in people's attitude toward what is valuable, a change American commercial culture could not endure. But the sugar maples hang in the balance, and we in fact decide their fate by the way in which we lead our daily lives. We can choose either endless economic growth as measured by increasingly questionable compensations for our frustrations (such as cars that come loaded with more diverting options to make endless commutes endurable), or we can choose a healthy connection with convivial natural surroundings. We can't have both. To believe that we can routinely drive great distances to reach an unspoiled landscape full of natural amenities at the end of our journey is a fatal delusion. If we want to live in towns with healthy sugar maples, we simply have to live in them more and drive in and out of them less. We cannot out-commute suburban sprawl.

There is a sign of hope. The sugar maples planted along the roads a century ago achieved a kind of victory before they died. Year after year, they sent thousands of seeds spinning downwind into the adjacent forests, where they sprang up in great profusion. Sugar maple seedlings are shade tolerant and seem to thrive under the oak woods of Weston, especially on moist glacial till soils. There are several woodlots in town where sugar maple is now a significant component of the forest understory and is pushing its way into the canopy. This tree is indeed well suited to Weston's climate and soil and has been slipped back into our forest by European invaders after the Native Americans kept it burned out for thousands of years—an ironic twist of ecological fate, but we'll take it. If they can only survive a few centuries of anthropogenic atmospheric uncertainty, there will be maples in plenty to tap in Weston once again.

ECOLOGY

Making maple syrup is a sustainable use of the forest. Because it doesn't kill the tree it is immediately appealing, like shearing a sheep or milking a cow. Careful, moderate tapping has no long-term ill effect on a maple tree.[3] The wound heals over quickly, and as the tree grows, fresh sapwood is laid down over the scars of old holes, which gradually recede into the heartwood. A sugar maple can be safely tapped when it reaches about a foot in diameter and throughout its life thereafter, which in a healthy environment may surpass a century and a half.

Just as wool and milk are improved by careful tending of the animals, so sap from a grove of maples is much improved if the trees are well cared for.

The prudent sugarmaker needs to be looking after the next generation of maple tappers and ultimately the next generation of maples, too. Maples need room to spread their crowns in order to produce the most sap, which is one reason roadside trees are ideal for tapping, unless they are dead. Maples in the forest do best if well thinned. Young trees need to be planted, or natural regeneration nurtured to replace those that grow old and die.

Planting of maples along the roads in Weston is useless. Salt kills the young trees even faster than the old ones. Some towns have done better. In neighboring Lincoln, the tree warden, Russell Barnes, began noticing distress in a double row of sugar maples along Baker Bridge Road near Walden Pond, back in the 1970s. He protected a few selected seedlings behind the stone walls, on the edge of the adjacent fields. About the same time, Lincoln greatly reduced the use of salt on its roads. Russell's young trees have done him proud, and Baker Bridge Road will soon have magnificent maples again. I like to cycle down that road and reflect on the wisdom of a little foresight and forbearance.

In Weston, our strategy is not so much to plant trees ourselves as to take advantage of the natural spread of sugar maple seedlings that has been going on for decades. We would not want maples to take over our forest, but we can encourage them in a few choice spots that have been selected for us by nature and chance. For a century, seeds floating on the southwest wind have sown a new generation of sugar maples that now range in size from saplings to tall trees over a foot in diameter on both sides of a little valley east of Weston Center. The maples are so thick that on bright autumn days the valley slopes look like a displaced corner of Vermont, full of luminous gold. For several years now, the Conservation Commission and Land's Sake have been thinning the woods along the stream. The object is not to create a pure maple stand—there aren't enough maples to do that in any case—but to give the nod to thriving young maples where we can, easing them a little more quickly into the canopy of older oaks. We have begun hanging a few buckets on the larger trees. If we manage the woods carefully, we will someday have several genuine sugar bushes to tap in Weston, and they will be safely removed from the traffic. In time, we will restore maple syruping in Weston to its former glory.

ECONOMY

The economics of making maple syrup in Weston are encouraging, at least in the long run. People in Weston will pay any price for a pint of maple syrup made by their own middle-school children, regardless of the quality. They would take out a third mortgage on their second home, if necessary. Every spring on a Saturday in late March we put a sign out by the road that says, "Sugar on Snow," fire up the evaporator for the last time, and hold a traditional sugaring-off party. The kids make "leather aprons" from thickened syrup poured onto snow (or crushed ice, if snow is scarce), and everyone eats this maple candy with donuts and pickles—as I say, it's traditional. Folks buy out the entire season's production of four hundred or so pints of syrup in a few hours. I suppose in theory we could charge $25 a pint and make a handsome profit. As it is, we charge about what you would pay to order maple syrup by mail from Weston, Vermont—not cheap. No one complains about the price. But even given the high price, with so few buckets to draw on for sap the project cannot cover its costs on syrup sales alone. We no longer make any pretense of running a commercial mapling operation in the way we once did.

When Land's Sake absorbed Green Power and took over the maple project in 1991, we faced reality and traded in the old evaporator for one less than half its size. This streamlined the operation, although with our bucket count so diminished we are still woefully undertapped. We now augment our maple income by charging for the many school groups who come from other towns to tour the sugarhouse, and the Conservation Commission gives us a small subsidy for providing an educational service to Weston's young people. This is all quite sensible but disheartening in a way—it is a far cry from the heady days when we were hoping to hang three thousand buckets, boil like crazy for six weeks on end, and let the educational side of the enterprise look after itself. Of course, we were never going to run the world's most efficient operation by employing work crews of manic twelve- and thirteen-year-olds—the involvement of the kids was always the main point, not profitability. Still, the operation was big enough to be economically viable, which, as with most Land's Sake projects, was its great educational strength. We have been forced by circumstances beyond our control to run a smaller demonstration maple project for now, and meanwhile to work with nature to bring sugar maples back to the woods.

EDUCATION

The educational value of something like a working sugarhouse is incalculable. Every suburban middle school should have one. This is the kind of thing that kids usually see only on field trips or, worse yet, in a museum: "Here's the quaint way we used to live, boys and girls." In Weston, the evaporator is outside the school door, just a few steps from the spot where the buses pull up at the end of the day. And when we're operating, the kids are right in there skimming scum from the back pan, helping stoke the fire, and pulling a stream of bubbling syrup into a bucket. They regularly "forget" to catch the late bus home and have to call their parents to come collect them for dinner, but I seldom see a mother or father who looks annoyed about this. They are overjoyed to see their preadolescent youngsters so turned on by anything. The children are involved in the whole process of making syrup from beginning to end: drilling holes in the trees, collecting sap, splitting wood, bottling the final product—which is, of course, a universally sought-for delicacy. There is nothing like making maple syrup to give you a feeling of indisputable accomplishment. The sugarhouse reeks of positive reinforcement.

The maple season starts with tapping out on a Saturday early in February. We meet at the sugarhouse and load a few pickups, trailers, and vans with galvanized buckets, lids, taps, hammers, and a couple of old braces fitted with 7/32" auger bits. Each tapping crew consists of a swarm of eager novices and a few calmer old hands, heads of families whose children have been involved for years and who still turn out for the fun of it. We fan out to the few pockets of healthy trees left in Weston. Some of the best maples survive in the cemetery, firmly rooted in the remains of the people who planted them around town a century ago. This provides a ready-made parable that I inflict on a new bunch of tappers every year as we blithely hang buckets among the headstones. Even in death, those who planted the maples care for them better than we do.

Once the buckets have been hung, on days when the sap is running we pick up a crew at the sugarhouse after school and make a collecting trip around the town. When we reach a row of trees, the kids deploy with their white collecting pails, while the driver climbs up in the bed of the truck. The collectors pour the sap (clear as water and only slightly sweet) from the buckets into their pails and head back to the truck. Some of them can't keep the pail off the ground with one arm and have to carry it with two hands, swaying from side to side as they walk. They hoist the pails, and we pour them through the first of many filters into the collecting tank. Back at the sugarhouse the sap is pumped through another filter into one of several holding tanks along the north and east sides of the building, out of the tainting sun.

When we have enough sap on hand, we boil. The thickening sap progresses from the back pan, where most of the boiling takes place, through a series of baffles in the front pan until it is drawn off as syrup in small batches from the finishing trough. We measure the specific gravity of the syrup with a hydrometer on its way into the bucket. The syrup is poured hot through a thick felt strainer into a tank, from which it is bottled. In recent years we have taken to drawing the syrup off a bit thin and finishing it on a more controllable propane unit. It takes about forty gallons of sap to produce one gallon of syrup.

In the belly of the evaporator is a fire that constantly needs to be fed. We burn mostly cull pine from our forest operation, pitching it in at a furious rate. For years I had bare spots where the hair was missing on both of my wrists, just beyond the back edge of my gloves, from swinging open the cast-iron doors and tending that fire. We aim for a full rolling boil across both pans, fine golden bubbles in the front pan and large white ones bursting in the back. The sugarhouse fills with sweet-smelling steam, until everything is dripping with it and caked with a microscopic layer of sugar. Meanwhile, outside this intense, contained world of fire and bubbling sap could be a sea of mud in a gray drizzle, or a howling northeast blizzard, or a crisp winter night, or a startlingly bright, clear spring afternoon. I have boiled under all these conditions and found that one can change to another without my noticing it, so that I am startled when I step out the door into a world different from the one I expected. I love boiling for long hours, then escaping for a few hours into the bright outdoors with a collecting crew, then plunging back refreshed into the alchemical sweatshop of the sugarhouse to make more golden elixir.

The middle-school students rush in after school or even during their study halls and lunch breaks, throw off their coats, and take up some task at which they have established themselves as seasoned professionals. They haul in wood, check levels, skim scum, help move pumps, check levels, bottle syrup, fill the water heater, check levels, wring out filter cloths.

"Hey, it's below an inch!" an urgent and unseen voice shouts through the steam from behind the back pan. If the evaporator runs too low, the syrup will scorch.

"How *far* below an inch?" Maybe we verify the alarm and adjust the arm on the surging float a notch to let in more sap. Those students who have proven themselves to be steady, reliable hands are initiated into more demanding jobs like stoking the fire (one side at a time and work fast, to keep the front pan from cooling) and drawing off syrup, the most privileged task of all. We show them tricks like waving a knife with the thinnest smear of margarine on it through the top of the back boil when it threatens to spill over the walls of the pan, which causes the boil to subside instantly to a sheet of deeply troubled sap. Then we spend weeks trying to keep them from working this spell unnecessarily, which frustrates the boil too much. No team of educational specialists could possibly contrive a better activity for middle-school students than making maple syrup.

Once or twice a season a heavy sap run continues into the weekend, and we declare a night boil. Now the old guard takes over, guys like Bear Burnes and Ned Rossiter, who helped McElwain get the maple project started over twenty years ago. Sometimes Bill himself sticks his head in the door to ask, "How's it going, lads?" Once a year, give or take, we assemble to chew the fat, talk Weston politics, reminisce about the old Green Power days, and boil syrup far into the night, as long as we have sap ahead. We could be almost anywhere in up-country New England, any year in the past century. Every now and then we concentrate our attention and draw off another batch of syrup. I sit on an overturned white bucket by the firebox door between bouts of activity, my brain steamed down to the size of a walnut inside my maple-crusted skull, trying to remember what it was like outside when I lit the fire that morning. And remind myself that I still have to make a late lamb check on my way home, but that tomorrow, Faith will look after all of that while I sleep.

As the March night chills and deepens, Ned goes out into the stiffening mud and moves the pump from tank to tank, systematically draining the sap out of them. Bear bottles syrup and washes filter cloths, sponging down and straightening up the operational rubble of the long day. I love working with these guys not just for the company, but because we have been at it for so long now—long enough that it has gone beyond being a neat idea that lasted a little while to becoming part of our lives, a part of the history of the town for an entire generation. Children, including theirs, have now grown up in Weston with syruping, cidermaking, and all the rest of our artfully contrived program of rural activity as part of their world. Making syrup with these guys reminds me that even if we were to stop now, we have already succeeded.

When the sap is down to the last hundred gallons or so we quit stoking, swing the firebox doors open to let the fire burn down, rake out the coals, pull the plug on the cupola fan, pack the jugs of syrup into my truck to take to the office for safekeeping until sugaring-off day, and head home under the early morning stars.

ESTHETICS

Driving home through Lincoln a few years ago, I stopped just north of the Weston line and took a walk into a small piece of woods I had not visited before. A friend had told me there were sugar maples in a little valley there, so I thought I would take a look. It was a bright, clear February morning, temperature in the midthirties, a little snow on the ground—just getting warm enough for the sap to run. As I headed up a trail past a series of ancient millponds I was surprised to find I was walking through a grove of dozens of mature, soaring sugar maples, many two feet in diameter. They stood not in rows but scattered naturally among other hardwoods—they must have been the offspring of trees that flourished back in the middle part of the nineteenth century, long before most of our roadside trees were planted. Perhaps a single maple tree growing in a yeoman's dooryard had quietly transformed an entire woodlot. The rolling landscape, the healthy silver-barked maples, and the sharp spring morning made a scene so perfect for tapping that I was seized with the idea of hanging buckets on those trees, fantasizing about it the way I did when I first read Helen and Scott Nearing's *Maple Sugar Book* as a teenager. It took me a minute to stop and remember: "Oh, right: I *am* making maple syrup. We hung our buckets a couple weeks ago. We'll probably be boiling tomorrow, if this weather holds. This isn't something I just dream about. This is something I do."

Part of my elation no doubt came from discovering a treasure so close to home; I wasn't much more than a mile from my front door. Partly it was the perfect morning, and partly it was the joy of finding maples so healthy and flourishing—they gave me hope. But mainly I was elated because the idea of making maple syrup is one of the most innately exciting things about living in the northern part of America. Boiling maple syrup is a distillation of place. It taps into one obscure quirk in the biology of forest trees in northeastern America—the clear, untainted sap that sugar maples run up and down inside their bark for a month or two before actually swelling their buds. It transforms this quality, boils it down into a sweet liquor that tastes like nothing else on the planet. The activity and the outcome are distinctly regional, a marker of home.

But beyond capturing a part of the world, maple syrup distills a quicksilver season. Syruping celebrates not so much the arrival as the less tangible anticipation of spring. That is the source of the fleeting flavor found only in the earliest syrup, not the heavy, dark stuff made later in the year. The days are growing longer as we faithfully hang the buckets every February, but because of the ponderous thermal lag of the planet, several more bouts of cold weather and heavy snowstorms are ahead of us. So are the seemingly endless spells of chill rain and drifting fog and the infamous weeks of mud as the frost grudgingly relaxes its grip. None of this backsliding deters the maples. They are deeply aware of where the world is bound to go. They draw their confidence from the knowledge stored in their roots of where the world has been, the concentrated experience of the previous growing season, the enlightenment their leaves have absorbed. Just as honey is made from the nectar that flowers provide, allowing bees to survive the winter and pollinate more flowers, and just as hay cut during the summer feeds the ewes in winter so they can raise their lambs to eat more grass, so maple syrup is the condensed connection between one year and the next. It is the first stirring of the new year, and sugarmakers catch it just as it rises.

I suppose to some this sounds like so much romantic treacle. That does not make it any less satisfying to those of us lucky enough to be caught up in it, year after year. Reading about maple syruping is a pale pleasure compared to picking up a brace and bit and trying to figure out where to put a hole in a tree. Any bunch of suburbanites can get together and make syrup—the only requirement is healthy maples. Some may object that the trees are so threatened we dare not tap them; for the last few grand maples by the town green, they may be right. But that is tokenism. The truth is, if we do not change our ways, those last trees will soon die anyway. If more people don't start *making syrup,* instead of just driving around in the fall to gawk at the foliage and buying corn syrup with imitation maple flavor, we don't have a prayer of making the changes necessary to save the maples from destruction. The health of maples cannot be merely preserved—it must be passionately embraced by people with a deeply felt stake in the matter.

Every February modern New Englanders complain about the late winter blahs and how the snow has worn out its welcome. Everybody seems to be reading from the same tired old script. Television ads for escapes to Florida and the islands appear. People burn up jetloads of fuel to get away from—what? The dullest, dreariest time of year? Sap is rising, lambs are dropping, garden seedlings are poking up in the greenhouse! By March we at Land's Sake seem to inhabit a world completely different from that of our winter-weary neighbors, who gladly shell out a couple of thousand bucks to flee the liveliest time of year. We are manic, they are numb. The quickening of spring is as dead to them as the road slush frozen in their wheel wells. Only that day dawns to which we are awake, said Thoreau, and I say only that spring comes to which we are engaged. Our culture is coasting deeper and deeper into neutral slumber, cruising through the living world with the automatic windows rolled up tight. People are no longer awake even to the everyday morning before their very eyes, let alone to any higher dawn. The bright spring sun puts out eyes attuned only to video screens.

The year we pulled out the old evaporator and put in the smaller one, I was boiling alone one Saturday afternoon when a woman in her late twenties walked into the sugarhouse with a little girl in tow. "Wow! I can't believe this is still here," she exclaimed. "I used to work here back when I was in junior high."

"Really? When was that?"

"Back in the seventies. I remember Bill, what was his name, again? Bill McElroy? The guy with the green sweater and the cigar. He was really cool."

"McElwain."

"Yeah. Is he still around?" she asked.

"Bill? Sure. He looks in every now and then."

"Wow, I can't believe it. It's all still here," she repeated. "We just bought a house in Weston," she added, a bit uncertainly, as if that, too, was a bit hard for her to believe. It is unusual for a native to return to Weston.

"That's nice." I turned to stoke the fire, and she squatted down by her child to watch, beneath the steam.

"Oh, my God!" I heard her laugh. I looked around. She was staring at the evaporator with her mouth open.

"What is it?" I inquired.

"It looked so much bigger when I was little!" she exclaimed.

"Isn't it amazing how growing up can change your perspective on things?" I agreed.

All right, the evaporator *was* bigger back then. Eventually I admitted that it wasn't just that she had grown, but that the evaporator had shrunk. It had shrunk because we, the citizens of Weston, collectively killed our trees. Perhaps in another decade her daughter will help us make syrup from new trees, trees growing back in the woods away from the deadly cars. And by the time her *granddaughter* helps make syrup, if we take good care of the forest, perhaps we will need a big evaporator again and have healthy trees to feed it.

Apple Cider

The cider press is the oldest piece of equipment that Land's Sake owns. It is the same press I encountered on the first day I worked for Green Power, although since then we have set its iron in new oak. We have no idea when it was built, but the patent stamped on its cast-iron grinder says 1872. The cider press has a magical appeal. We set up shop beside the farmstand on weekends in the fall, and the customers and their children swarm around the press like bees. Amateur cidermakers, they drop in the apples, turn the crank on the grinder, and screw down the plate that squeezes out the juice. People will cheerfully pay extra for cider they themselves just made. Most days, our part in making cider amounts to carrying boxes of washed apples to the press, lugging the pressed barrels away to the pomace trailer, pouring tubs of cider into the bottling tank, and selling cider to the very people who are happily grinding and pressing it.

The first fall that I made cider at Green Power, in 1975, the apples we pressed came from abandoned orchards on Applecrest Road and Blossom Lane. Many of the apples were Baldwins, a large, tart New England apple that gave the cider we made its exceptional flavor. The bumper crop we found under the abandoned trees was a fluke, for the conditions needed to produce such a bonanza occur not more than once a decade. Only 1992 has compared to that year. That is a long time to wait to make cider. Other years we went to commercial orchards in towns ten or twenty miles farther west and bought drops by the bin. Occasionally we found flavorful varieties like Baldwins or Russetts, but mostly they were McIntosh, Macoun, Cortland, and Delicious, insipid modern apples that made cider too bland and too sweet for my taste. The selection became more and more limited as the years went by, and growers steadily replaced their last blocks of Baldwins. We needed our own source of cider apples.

For years we tried to create such an orchard in Weston. In the 1970s there was a neglected orchard of two or three acres on the Weston College land, opposite our cornfield on Concord Road. It occupied what appeared to be a perfect site for apples, a slope that rolled down to a belt of red and white pines by the shore of the college pond. A few times every winter we took a break from the sugarhouse and devoted a few days' work to the orchard. By the end of March, the weather softened, and we pruned into the twilight as the returning redwing blackbirds sang by the pond. You couldn't have asked for more satisfying work, or any that held more promise for the future, than tending apple trees.

The only problem was, nothing we did made those old trees bear fruit. They bloomed faithfully, and we found a beekeeper to put a few hives in the orchard to ensure pollination. But most of the swelling apples dropped off the trees in June, while still tiny, and the few that did survive turned out small and gnarly. By late summer, the leaves grew yellow and splotched. We dutifully fertilized the trees and borrowed an antique sprayer to coat them with Superior dormant oil every May, when the buds were between silver tip and half-inch green. We hung white squares coated with tanglefoot to trap European apple sawfly and tarnished plant bug, and sticky red spheres to trap apple maggot. It was not enough to revive those trees. Among any number of other problems, a common fungus called apple scab infested that old orchard, dooming our efforts to failure.

Land's Sake contracted with the Conservation Commission to systematically renovate the orchard. We planted young trees on semidwarfing rootstocks, filling gaps where older trees had died or had been removed because they had grown too sickly. Our strategy was to replace the old orchard stealthily so as not to shock the people who liked to walk there. Most of the new trees we planted were actually old trees—antique New England varieties that were reputed to have some tolerance for the scab fungus that is so debilitating to many modern commercial varieties, and that would provide both good eating and a superb cider mix—Baldwins and Russetts, Rhode Island Greenings and Yellow Newtowns. We also planted several recently developed scab-resistant varieties, although they were said to bear uninspiring fruit. I even set aside my organic scruples and sprayed the young trees with fungicides, using a backpack sprayer. We began the new orchard on the site of the old with high hopes and with what seemed to be a good compromise "integrated pest management" strategy for raising a crop that was widely acknowledged to be next to impossible to grow organically.

Still the trees did not flourish. After a few years we decided it was asking too much to expect the young trees to thrive with the old ones still in place, constantly showering them with scab spores. Sadly, we removed all the old trees and filled out the new orchard with scab-tolerant Baldwins and Russetts, along with a few Northern Spies, which have no resistance to scab but are wonderful apples. I continued spraying. I got out in the orchard within a few hours of the beginning of every scab infection period—in other words, in the midst of every spring rainstorm—thoroughly coating the leaves and blossoms with fungicide to be redistributed by the rain. I did not enjoy it, but I did it.

In spite of everything we tried, the trees did not grow robustly or come into bearing as hoped. They made particularly poor growth during the dry summers that marked the late eighties and always looked rather spindly and sad. Many died when pine voles chewed off the roots beneath their plastic mouseguards, which we discovered when we pulled the slumping trees out of the ground as dry, dead sticks one spring. Others were girdled high on their trunks before we could stamp down the deep snow that had drifted over the guards. Doug Henderson and I exhausted ourselves in an hour one glistening January morning, tromping around to every little tree, breaking through multiple layers of snow and ice two feet deep, capped by several inches of solid crust from the latest storm. It was all to no avail. Every year something else would go wrong, and the relentless scab never did come under control. Tired of using sprays I didn't like and convinced the orchard could never succeed on a minimal spray schedule, I advised the disappointed Conservation Commission that we should abandon the orchard project. They were determined not to throw away their investment, however, and decided to press on with a regime of heavier chemical spraying. I turned the orchard over to others on the Land's Sake staff who were willing to try this solution. But our apple trees have never yielded more than a meager harvest.

The refusal of the college orchard to respond to my care and bear fruit is the greatest failure of my agricultural career in Weston. I take this very personally. For some reason I have been able to do everything from growing strawberries to managing woodlots with modest success, but not to make a little piece of the earth say apples instead of scab. Most of the trees we so lovingly planted have long since died; those that survived rarely produce fruit, and the fruit they do make is scanty and misshapen. This failure is galling to me because there is nothing in the world for which I would rather be remembered than as an innovative orchardist, and instead I will be lucky if my orchard career is simply forgotten. Why is this? The most likely explanation is that certain skills, like working wood, fixing engines, keeping bees, and growing fruit trees, require a kind of quiet patience and light touch that I lack. The explanation I prefer is that it is all a consequence of ecological history, a subject requiring mere doggedness that I *have* mastered and can exercise to my own advantage. The part of the world where I live has been made inhospitable for the kind of apples I want to grow, by forces of cultural myopia beyond my control.

The apple tree is not native to America, but it was a fortuitous introduction and adapted well to the New World—or so it seemed for two centuries, at least. Apples arrived in Massachusetts Bay along with the English in the 1630s and quickly flourished and spread. The soil and climate suited the trees, and they were widely planted in every town. Apples were grown primarily to produce cider (which we call hard cider, to distinguish it from the unfermented sweet cider more commonly consumed today), and only secondarily for fruit to be cooked in sauces and pies or eaten fresh. The trees took so well that within a few generations cider had replaced beer as the everyday beverage of New England. Prodigious quantities were consumed. It was common for an eighteenth-century Massachusetts farmhouse to have six or eight barrels of cider in the cellar, and per capita consumption rose to well over a barrel for every man, woman, and child. In truth, most of it was drunk by men, who also drank a good measure of rum.[4]

The adoption of cider was a dietary change with ecological roots. Back in England, cider was enjoyed in a few districts, but the English were mostly confirmed beer drinkers, which they remain today. Beer is brewed from barley, of course, and barley was the ordinary, workaday porridge and beverage grain of the English countryside. Growing barley presented real problems in New England, however. Neither wheat nor barley did well in the dry, sandy soils that were most commonly tilled in eastern Massachusetts, and farmers here quickly settled on Indian corn and English rye as their principal bread crops. Barley was malted and beer was brewed, but unlike rye, barley required manure to make a decent crop. Manure was in short supply and was usually reserved for the Indian corn. Then, too, barley had to be sheared during the short summer season, squeezed in between reaping the rye and haying the meadows during the hectic months of unrelenting toil, in an era when small grains and grass were both mowed by hand. In effect, insistence on beer reduced the yield of the maize crop by diverting manure and curtailed the acreage of rye and hay a farmer and his sons could harvest by diverting labor. It was enough to make these stalwart Englishmen, the sweat dripping from their beetled brows, think twice about old John Barleycorn—maybe as the ancient ballad suggested, he *should* die, and this time stay dead for a change.[5]

Apple cider solved these problems and, in a sense, completed the transformation of English husbandmen into *New* England farmers. Apple trees could be grown on marginal, rock-marbled slopes—they did not require good tillage land. Farmers discovered that orchards, like pastures (and sugar maples for that matter), did best on hillsides composed of glacial till, where the blossoms escaped late spring frosts and the roots could find plentiful summer moisture oozing along above the boulder clay. Apples, their perennial roots reaching into the mineral-rich subsoil, did not require annual manuring. Best of all, apples ripened in the autumn after the other crops had been safely gathered in and there was time to make cider. All in all, apple cider fit the soils and seasons of New England much better than beer.

But this story leaves an ecological mystery unsolved. Today, as I have learned from bitter experience, it is next to impossible to grow apples in New England by organic methods—even if the legions of insect pests can be drawn to their doom by sticky spheres and pheromone traps, apple scab fungus still debilitates the trees and destroys the crop nine years out of ten. Yet every colonial yeoman had his orchard of unsprayed cider apples that apparently yielded reliable crops of twenty barrels of cider per acre. This six hundred gallons or so represented perhaps three hundred bushels of apples—not a huge yield but decent. How did they do it?

Perhaps apples did well by escaping many of their traditional pests in a new land, just as rubber plantations flourish today in Southeast Asia but not in their native Brazil. It took some time for Old World insects such as coddling moth to follow their host to America, and for native American insects like plum curculio to discover domestic apples and become pests. This may partly explain apples' early success, but many fruit scientists believe scab has been in New England almost as long as the apple itself. So how did cider become the everyday beverage of pesticide-deprived husbandmen when today it cannot be grown without extraordinary chemical protection?

The apples being grown for cider in colonial New England were probably resistant to the strains of scab that were in the air at that time, and so yielded a decent crop. Certainly the trees and possibly the infecting fungus itself have changed since then. In early America, cider was not made from grafted varieties of apples but mostly from seedling trees. Apples do not breed true to type. The seeds from a Red Delicious apple, for example, will not produce Red Delicious apple trees (thank goodness), but a wide range of mostly small, sour fruit. Not many of them will be good to eat, but they will make good cider. New England farmers planted apple seeds in the back corners of their gardens (or, as Thoreau suggests, simply looked through what sprouted from their pomace heaps) and chose the most vigorous seedlings to set out in their orchards. By constant selection they created what plant geneticists call a land race, that is, a population hardy and well adapted to local conditions and resistant to such common diseases as scab. If most of the fruit was marred by a few scars and blemishes at the end of the season, it made little difference to what flowed from the press.

In every orchard of a few dozen trees, by chance there would have been a few whose fruit was good enough to be packed away in barrels, providing winter apples for months to come. Scions from the best varieties circulated locally, and a few found wider fame and were grafted throughout the region. The Roxbury Russet appeared in that town (now part of inner-city Boston) in the 1640s, while, according to legend, the Baldwin was discovered in a Woburn fencerow by the engineer Laommi Baldwin, out surveying the line of the Middlesex Canal in 1793.

The nicely balanced system supplied, first, a distinctive drink and, second, a wide variety of cooking and table fruit. It was undercut not by any ecological flaw, as far as I know, but by an outbreak of moral rectitude. Early in the nineteenth century, the temperance movement swept through New England. Even mildly alcoholic cider was condemned from the pulpit, and coffee and tea became the most common daily drinks. This was part of the same wave of cultural reform that brought about a more commercial approach to farming in general, as discussed in the previous chapter. It was no longer acceptable for a Yankee farmer to go about his business half-buzzed—he was now expected to be a coolheaded calculator who kept close accounts. A barrel of cider had no place in the cellar of a farmer aspiring to climb up among the sober middle class. In the 1830s and 1840s many orchards were chopped down in a fit of temperate zeal, while others were simply abandoned and resigned to pasture. Hard cider did not vanish from New England farms entirely, but consumption slumped to negligible levels. The demise of "vast, straggling cider orchards" full of zesty apples was publicly mourned only by the hard-bitten Thoreau, who lamented not so much the cider as the lost "tang and smack" of the feral, ungrafted fruit itself. If Thoreau found the Baldwin insipid, what would he make of the Red Delicious?[6]

Even as cider orchards disappeared, they were being replaced by orderly ranks of grafted trees. A market for fresh table fruit was developing in nearby cities and milltowns, and forward-looking farmers in Middlesex County raced to meet it. This second orchard era in New England sprang up over the stumps of the first and flourished for another century before starting to wither, though it has not quite perished yet. The new fruit crops were planted not to contribute to local sufficiency in food and drink, as before, but strictly for sale. Orchards of grafted apple trees producing fresh fruit represented a larger investment and were generally more carefully managed than their predecessors. Cider became a mere by-product of the apple business, as it remains today. A premium was placed on unblemished, cosmetically appealing fruit. By the end of the nineteenth century, professional orchardists were beginning to spray their apples. This brought about a radical transformation in orchard ecology.

To combat such familiar insect pests as tent caterpillar, canker, plum curculio, apple maggot, and coddling moth, progressive farmers applied botanical poisons like pyroxins and ryanines, and "arsenial" compounds such as Paris Green and London Purple. Paris Green, a form of copper arsenite used to color paint, began to be used as an insecticide in the 1860s. London Purple, a calcium arsenite by-product of dye manufacture, came into vogue about 1880 because it was cheaper and adhered better to leaves and fruit. It was also found to injure the foliage, however. In 1892, less damaging lead arsenate was developed by a Massachusetts state researcher named Francis

Moulton in the fight against the gypsy moth, which was ravaging local oak forests. Lead arsenate was effective against many pests and remained the most widely used orchard insecticide until World War II. To combat such fungal diseases as rust and scab, growers relied chiefly on Bordeaux mixture, compounded of copper sulfate and lime.

A few chuckleheads worried that Paris Green et al. might harm the people who ate the fruit or poison the cattle that ate the grass under the trees, but in 1899 former Secretary of Agriculture William Sessions declared, "This fear of Paris Green for cattle is all moonshine and nonsense." We have heard that kind of comforting talk from our agricultural leaders before, or I should say, since. After World War II, the heavy metal and botanical pesticides were replaced by a host of more sophisticated synthesized chemical sprays, led by DDT. To what extent fretting about the health effects of these biologically novel poisons is all moonshine and nonsense I will leave to the reader's own, no doubt already settled opinion. But human health risks aside, these sprays have made it virtually impossible to grow apples organically.[7]

There are a number of reasons for this. For one thing, spraying made it possible to grow varieties of apples that are highly scab-susceptible, for example, McIntosh and Delicious. These now dominate the market and the landscape. Also, as is now well understood, the spraying of pesticides suppressed populations of many beneficial insects in the orchard. This allowed outbreaks of previously inconsequential organisms such as red mites, elevating them to pest status. Helpful fungivores that once preyed on apple scab have also been widely eliminated. But you would think that by planting the more resistant of the older varieties and allowing a more balanced orchard ecology to reestablish itself in the absence of chemical sprays, we could at least produce good yields of cider grade fruit, if not cosmetically perfect table apples. Perhaps in time that will be possible, but our experience at Land's Sake has been that such trees barely grow and scarcely bear. Some apple scientists wonder whether a century of fungicide spraying may have selected for scab fungus that is not only more resistant to sprays, but more virulent in its effect on the trees. Research on this subject is in its infancy. It may be that my clumsy attempt at spraying merely suppressed the milder strains of scab fungus and allowed the meaner ones to dominate in our small orchard. All in all, we seem to have created conditions that will frustrate attempts to grow apples organically again in New England for a long time to come.[8]

The apple industry in Massachusetts has packed itself in a tight box. When the era of commercial orchards began a century and a half ago, there were hundreds if not thousands of named apple varieties in circulation, and many more appeared over the following decades. A wide range of apples was grown for market, and a still wider range survived in smaller farm orchards. But as the industry developed in the twentieth century, the number of commercial varieties steadily narrowed to those few that stored longest, shipped farthest, and had the most uniform, glossy appearance. As the varieties being grown shrank, Massachusetts growers came in for increasing competition from a national and then international fruit market that flooded the supermarket shelves with apples. Apple production in Massachusetts reached its height in 1900 at three million bushels and then held its own through World War II. Since 1950 it has declined to about two million bushels, giving way to apples from Washington, California, and beyond. Smaller, more intensively managed trees have allowed production to be concentrated onto many fewer acres—there were more than 40,000 acres of orchards in Massachusetts in 1930 compared to about 7,000 today.

The geographical center of apple country has moved steadily west across the state in the face of suburban development and rising land values. Production in Middlesex County, once the heart of the orchard region, has dropped from about a third of the apples grown in the state to about an eighth. Orchard acreage in the county is down from well over 10,000 to fewer than 1,000 acres. Sadly, the orchards are nearly gone from my part of the world. Weston had 269 acres of apples in 1875—you could probably see an orchard from almost any place in town. Today, there are next to none.

The apple growers left in Massachusetts, in direct competition with growers all over the world, struggle valiantly to produce a few varieties of cosmetically perfect fruit to sell in a glutted market. They battle to grow highly pest-susceptible apples in a moist, moldy climate that is home to well-entrenched pests and high land values, while their competitors bask in the desert sun and irrigate with water provided by the government at subsidized rates. Local growers thus face high costs and low prices in marketing their quality fruit through established wholesale channels. If Massachusetts orchards are to survive, something needs to be done either to cut the cost of what goes into an orchard or to raise the price of what comes out, or both. Some progress has been made in the past two decades to reduce spraying through integrated pest management. These efforts have not cut production costs enough, however, to allow Massachusetts McIntosh and Macoun apples to bring more than a meager profit, while the tax cost of the land under the trees steadily rises. And so more orchards are inexorably driven out of business. Meanwhile, there is strong demand for cider, and not enough apples to fill it because cider is made exclusively from culls. Oddly, it seems orchards are spending too much money to grow the wrong kind of fruit.

ECOLOGY

I certainly can't suggest a simple solution to this dilemma. As a flop at growing apples myself, I am not about to prescribe for the entire industry. Yet I am certain that if Massachusetts apple growers cannot move away from such tortured ecology and toward selling of more distinctive regional products, they will surely vanish from the landscape. This is the march of progress that we know only too well. Of course, the greatest single threat facing orchards (and all other kinds of farming) is rising land values and development pressure, and the only response to that is a set of protection measures for agricultural land. But even if land is saved—and perhaps in order to make it more likely that orchard land *will* be saved—apple growing needs a new foundation upon which to grow and flourish. Organic cider orchards may eventually be a base for a more sustainable New England agroecology and economy.

The progress made over the past two decades in integrated pest management is encouraging, and continuing research should take us steadily toward biological controls for many pests. But a profitable apple orchard that is entirely organic is not going to be easy. Some pests, like apple maggot, for instance, can be controlled directly with traps, but others, like plum curculio, may prove less trap-prone. Early farmers didn't have to deal with plum curculio, which made the jump from native species to domestic apples only in the past century and now causes nearly complete "June drop" of developing fruit if uncontrolled. Scab presents the greatest challenge of all. We will need to rely heavily on scab-tolerant varieties of apples if fungicides are to be greatly reduced or eliminated. Perhaps in time a less harsh chemical environment and the resurgence of natural fungivores can alter orchard ecology in such a way that less virulent strains of scab prevail.

There are several ways to create scab-tolerant orchards, and they are by no means mutually exclusive. One is to plant varieties of scab-resistant apples that have been developed by fruit scientists over the past few decades and to continue developing more of these trees. Another is to plant a wide range of the older varieties that have at least some scab tolerance. A third, more radical approach is to begin growing nurseries of seedling trees and selecting the most vigorous to set out in orchards, once again creating a land race of apples that can thrive in the current pest environment. Such changes make little sense, however, unless a part of the orchard industry begins concentrating once again on producing cider primarily and fresh fruit secondarily.

The ecological advantages of growing apples primarily for cider instead of table fruit are obvious. First, the cosmetic quality of the fruit ceases to be important, so the level of spraying required can be greatly reduced. But the superiority of cider apples is more than skin deep. A major problem with dessert apples (as compared with many of the small fruits and vegetables that can be grown organically without much trouble) is that a fancy apple is a large, sweet, soft, thin-skinned object, attractive to almost everything with a mouth, that hangs there for months on end while it slowly ripens. Naturally, squadrons of pests riddle it from stem to calyx. Cider apples, on the other hand, are perfectly usable (in fact often preferable) if they are hard, sour, and even somewhat bitter—the very physical and chemical qualities developed by wild fruits to protect themselves. Take the Roxbury Russet—it isn't strictly a cider apple but it comes down from the cider era, and it is one tough customer. Concentrating more on cider would allow us to move back toward the ecologically sustainable base of a pest-resistant land race of sour apples. That land race, in turn, would provide a wide pool of genetic variation from which resistant table apple cultivars could occasionally be plucked, just as they were in the dim past before Paris Green. The continual discovery of special apples to eat would once again become a natural outgrowth of widespread local orcharding, a vernacular pastime rather than a specialty of professional fruit scientists and marketers. The rediscovery of diverse, hardy cider orchards may be one ecological direction in which we ought to be headed.

ECONOMY

How would such orchards fare economically? The market for sweet cider is large, but of course fruit to be crushed for cider pays much less to the grower than fruit to be eaten fresh. To make cider orchards profitable, production of run-of-the-mill fruit must be streamlined by reducing costs for spraying, picking, and storage. At the same time, the market for premium cider must be increased. Fruit juice drinks have been gaining steadily in popularity over the past decade, a hopeful sign. It would be nice to see apple cider push some of the synthetic, carbonated corn syrup soft drinks off the shelves of local supermarkets. A similar revolution from fresh consumption to juice has led to an enormous expansion in cranberry production in Massachusetts over the past few decades. Why couldn't apples follow a similar path?

There is a natural way to preserve apple cider for year-round consumption, needless to say, and that is by fermenting it. Hard cider is a pleasant, mildly alcoholic beverage that is making a small comeback in New England. There is no reason cider mills couldn't follow in the footsteps of microbreweries and recapture a part of the bloated beer market. Cider replaced beer in this part of the world for ecological reasons once before—perhaps it is time for this to happen a second time, three hundred years later.

Beyond hard cider there is apple wine, made by fortifying fermenting cider with sugar to bring it up to 11 or 12 percent alcohol. Over the past two decades, Nashoba Valley Winery in Bolton, Massachusetts, has been a pioneer in making high quality wines from apples and other New England fruits. They make several varietal dry apple wines, along with a tart cranberry-apple wine and an apple "cyser" (or honey mead) that is perfect for a cold winter night by the fire. Faith and I served Nashoba Valley's first batch of apple champagne at our wedding. What better way to add value to a distinctive local product and get many a marriage off to a good start, than to drink a toast both to and from home ground?

After sweet and hard cider, the third way in which organic cider orchards might succeed is by producing tasty antique table fruit. This would comprise a smaller part of the crop, just as a select part of the cranberry harvest is still handpicked for the fresh fruit market. Having low-maintenance cider orchards made up of seedling trees and scab-tolerant cultivars, growers could pick the best apples for fresh fruit sale and leave the rest to low-cost bulk harvesting for the cider bin. Organic orchards would be especially well positioned for the pick-your-own market, appealing to parents concerned about the safety of the fruit and the surroundings in which their children are picking. Both revived antique varieties and new land race–derived varieties should appeal to fruit fanciers looking for something more inspiring than the insipid fare that now dominates the market. To the Northern Spy and Westfield Seek-No-Further, we might add the Weston Spray-No-Longer and the Neopagan Pippin.

Cider orchards might well be an ideal niche for organic or near-organic apple growers in New England. Modern cider orchards might attain the ecological adaptiveness of colonial orchards, eliminating the need for spraying. The economic key, as always, will be to produce distinctive, high-quality products for a local market: premium sweet cider, hard cider, apple wine, cider vinegar, and dessert apples with distinctive flavor. These markets will have to be enlarged, but the recent success of new varieties of fruit, juice, and wine in the market is encouraging. Community farms could help develop this new old approach to apple growing by playing an educational role.

EDUCATION

Turning the crank of the old cider press is one of the best lessons that Land's Sake has to offer. The students recruit themselves spontaneously from the crowd that gathers around the press in the autumn sun when the clanking iron gears start up and the crunching chopper begins to bite into apples at the bottom of the hopper, spraying juice. Everybody wants a turn. Fathers help their small sons and daughters with the handle as it lifts their feet right off the ground. Mothers dangle their infants over the grinder so they can drop in apples and watch them get chomped and squished. The pomace spews into a wooden barrel sitting on a slatted rack beneath. When the

Farm visitor Annette Ramos, Ed Hagenstein, and Tom Gumbart press apple cider, 1993. Ed was Land's Sake's director at the time; Tom took over in 1994. (Neil Baumgarten)

barrel is full, we slide it down under the screw press and begin filling another one. Meanwhile, the full barrel is fitted with a heavy oak top that presses down when the massive cast-iron screw is turned, first by hand, then with a long four-by-four as the mash compresses. Cider floods out of the barrel and over the front lip of the press into a waiting tub, to a chorus of cheers from the crowd.

Within an hour, the whole sticky apparatus is crawling with soused honeybees and yellow jackets. Those who haul the pressed barrels off to the trailer and punch out the pomace plan to collect a few stings. Few people complain. This is real: apples in, cider out, a few bee-stings. A funky, clattering, comprehensible, hand-cranked machine of oak and iron, neatly engineered, grinds out a dozen gallons of cider every hour. People are thirsty for this kind of experience. All over the country, children's museums are springing up where kids play active, educational games of science. I have nothing against these places; they are good fun. But would we need them if we re-

Feeding the press.

tained some of the interesting reality of human-scale production in our daily lives? Believe me, nothing in a children's museum can beat an antique two-barrel cider press in action.

Pressing apple cider is spontaneous learning at its best, but orchards offer a larger opportunity for continuing education as well. People, particularly children, love picking apples and collecting drops for the cider bin. A few can also be counted on to come out to help plant trees. Older kids can be involved in pruning and in making, hanging, and monitoring insect traps. Over the years, we have engaged people with apples in all these ways at Green Power and Land's Sake, but because of the failure of our orchard to produce reliably we have never blended them in the comprehensive way we would like. Ideally, there is no better way to teach a fundamental lesson of ecology and history than through the complex story of cider orchards in New England; and no better places than the orchard and the cider mill to teach it.

Community farms are well suited to help revive ecologically sustainable orchards in the region because of the powerful educational role they can play. Private commercial orchards are heavily invested in land, trees, equipment, and storage facilities for conventional market production, and not many can reasonably be expected to experiment with growing antique varieties, let alone nameless seedling trees. Producing low-value fruit strictly for cider would quickly put these orchards out of business, given the large capital investment they represent. Community farms, on the other hand, occupy commonly owned land, so the high cost of land has already been removed from the economic equation. Diversified community farms such as Land's Sake can well afford to try out a few acres of cider apples, which require minimal spraying and storage facilities. If such experiments were repeated on community farms throughout the region, I think we could eventually find working methods for creating organic cider orchards. It may take some decades to develop a new race of scab-resistant cultivars that are also reasonably productive. But it took a few generations for the colonists who journeyed here from England to give up their customary beer and adopt cider in the first place. Ecological revolutions are not made overnight.

ESTHETICS

Apple orchards are beautiful. Beauty has value, and it might be more economical to keep it intact within our everyday surroundings than to spend money to procure it elsewhere. Orchards add seasonal color to the landscape with their white blossoms and red fruit and interesting winter texture with their twisted, angular trunks and limbs. They are at their best on hillsides, running up to the edge of the darker forest. In apple country the orchards often form belts along the valley slopes, between the hayfields below and the forest above. These are places where human intelligence has devised a landscape that makes sense, and stuck with it. What in the world possessed us to replace that sight almost everywhere with the trashy gaudiness of tract houses on Orchard Drive? If we are such a wealthy people, why should we tolerate a countryside bereft of orchards?

Thank goodness for the apple growers who hang on today and give us orchards to admire, and long may they wave. I am not going to tell them how to go about their business, but I do want to see more of them, not fewer, in the next generation. McIntosh and Delicious apples do not do justice to the landscape that produces them, and thus they cheapen and endanger it. They are a mass-produced, industrial fruit, and the taste of the approaching subdivision is already in them. The tart Baldwin makes a stronger statement, and the Roxbury Russet embodies an entire sermon on upright behavior. The cider that is expressed from apples like these is restorative in itself. A glass of pale, bone-dry apple wine helps us reflect on the way before us. A richer, sweeter glass of cyser with its spicy honey bouquet is a fit reward for a brisk January day spent pruning the orchard—or skiing in the orchard for that matter. More literally than any other natural product, apple wine distills an attractive part of the landscape to its essence. We may enjoy well-made wines from all over the world, but not in the same way as something drawn from our own backyard. To live in a New England restored to orchards, fine cider, and apple wine is a dream worth cherishing.

Cranberries

For years I dreamed of finding cranberries in Weston. I searched high and low for them. I say searched high because one day while still a student at Brandeis, before I even knew Green Power Farm existed, I was out walking in Weston and came upon a mat of cranberry-like plants covering a rocky outcrop on Doublet Hill. It was a vibrant evening in early May, the oak leaves just emerging. The little evergreen vines on the rock bore a profusion of small, waxy bells, diffusing an ambrosial aroma and singing with bees. I thought I had found the fabled mountain cranberry (*Vaccinium vitus-idaea*) and became ecstatic, as I frequently was in those days. Upon returning in the fall I discovered these were actually bearberries (*Arctostaphylos uva-ursi*), a distant cousin to the cranberry whose fruit is edible but dry and not particularly palatable. Bearberries are rare enough as far south as Weston; mountain cranberries, unheard of. Still, I continue to search high for cranberries because you never know what might turn up.

The common large cranberry (*Vaccinium macrocarpon*) of the lowly bog was supposed to be much easier to find in Weston than its rare mountain relative. Almost every farm in town, I was frequently assured, used to have a small bog tucked away in some back corner. Old-timers told me vaguely where this one or that one had been, back in Jericho Forest somewhere. Years passed; I soaked my boots in dozens of bogs and swamps all over Weston and never found a single cranberry plant, let alone tasted a berry. I began to think maybe there was something mythical about this fruit, too.

Finally, one day a man who had grown up in Weston told me about a small bog where he had picked cranberries as a boy. He pointed it out on the Forest and Trail map—it wasn't even that far back in the woods. This looked more promising. One afternoon in July, I left the trail and sought out that bog, pushing through the fringe of highbush blueberries along the shore and venturing onto the quaking surface of the bog. It was a typical sphagnum mat of an acre or two, overgrown with leatherleaf, pitcher plants poking up every so often. Before long, sure enough I began finding cranberry vines creeping here and there among the sphagnum and meadow grasses. I began to get excited. I started seeing a few hard, unripe berries, the fall crop on its way. And then I found something unexpected: a single, bright red, utterly ripe cranberry. It was like a little round sack of cranberry juice, perfectly preserved. It could only have been a fruit left over from the previous fall that had somehow lasted clear into the summer. This was a powerful portent indeed. I perceived that my luck with cranberries had changed.

I have returned to that bog many times. I have picked enough fruit to bring home in every month from October to March. Some years the water is so high that we can't get onto the bog to pick without wading, and other years snow covers the wild crop before we can pick it. But most years we manage to collect a few quarts, enough to store in the back of the refrigerator and dole out sparingly over the months. I like to eat them with Grape Nuts or dropped into pancakes before they are flipped or mixed with apples, walnuts, honey, and yogurt for lunch. The potent wild berries are too precious to be squandered.

Wild cranberries don't taste much like the cranberries you buy in the store. I was prepared to eat them with special relish because I had picked them myself, but I didn't expect them to actually taste much different from any other cranberry. The sour cranberry did not seem like the kind of fruit that could have subtleties of flavor. Store cranberries are red on the outside and white on the inside and are dry and crunchy. Cranberries in the wild bog are glossy, thin-skinned, and pure, bright red all the way through. And juicy: when you bite one it doesn't crunch, it pops! The flavor is acidic, but indescribably richer and sweeter than the store-bought variety.

After observing the berries on the bog for a few seasons it wasn't too hard to figure out what accounts for the dramatic improvement in flavor. Unlike cranberries from the store, which are sour the way green apples are sour and hard the way underripe peaches are hard, those in the bog are fully ripe. As the berries redden in September and October, frost begins to settle into the bogs on chilly nights, and all the berries that are not sufficiently ripe shrivel up and drop off the vines. Frost can take a substantial percentage of the crop, so commercial growers rake them well before frost strikes. The hard, unripe berries that are harvested early are also less prone to damage and easier to handle and ship than the soft, juicy berries they might have become. They are another industrial fruit.

Perfectly ripe berries, on the other hand, seem impervious to an astonishing range of cold and heat. They can hang there for months, as the first one I ever found did. We frequently pick them in quantity as late as December and even into the new year. Thoreau picked cranberries in Concord on November 19, 1853, and reported that "many of them being frost-bitten, they have now the pleasant taste of spring cranberries, which many prefer." Country people used to know these things. A Weston farm family named Coburn would first rake their cranberries for market and then go back later to pick for themselves. America may be rich, but as far as I know, it is virtually impossible to buy a ripe cranberry in this country. You have to go find it yourself.[9]

As my long quest suggests, though, cranberry bogs are hard to find. That every farm used to have one I can now categorically refute, but there were at least a few around. The bog where we pick was cultivated not that long ago. The drainage ditches are still visible, as is the dike at the downstream end that could be closed to flood the bog through the winter to control weeds and pests. Farmers used to cart loads of sand onto the ice and spread it out evenly, so that it would settle onto the bog in the spring and create a surface most favorable to cranberries. Today, most of these small farm bogs are overgrown by brush and trees and have succeeded to swamp. Only a handful have formed open sphagnum mats that still support a few feral cranberries.

At an earlier period in Middlesex County, cranberries flourished not as a highly cultivated crop but as a by-product of the mowing of the low-lying hay meadows every year. The meadows along the Concord and Sudbury rivers became famous for the cranberries that grew among their native grasses. The ripe berries were raked in the fall, long after the hay had been carted away. By the middle decades of the nineteenth century, many farmers reckoned the cranberries were worth more than the hay. Unfortunately, both crops were lost to flooding of the river meadows, caused primarily by the Middlesex Canal dam downstream in Billerica, but also by wholesale deforestation of the watershed, which increased summer runoff. The meadows grew up to sedges and buttonbush, and cranberries became less common and much less accessible. The town of Concord reported fifty acres of cranberries in 1855, only two by 1905. Meanwhile, many of the smaller meadows along tributary brooks in nearby upland towns like Weston had been thoroughly drained and converted to domesticated hay, as noted in chapter 3. Only a few boggy, hard-to-drain swamps and streamside meadows were consigned to cranberry production. Weston reported fourteen acres of cranberry bogs in 1865 and six acres by 1905. Like many other crops, cranberries in Massachusetts became a specialized business in the twentieth century.[10]

The domesticated cranberry and its culture developed on sandy Cape Cod and in neighboring Plymouth and Bristol counties, beginning in the 1820s. By the late nineteenth century commercial cranberry production in the state was heavily concentrated in that region and fading away to insignificance everywhere else. Total acreage doubled from about six thousand in the early decades of the twentieth century to about twelve thousand by the 1950s, where it has remained. Cranberry production held at about five hundred thousand barrels into the 1960s but has more than tripled since then with the advent of wet harvesting for the juice market. Cranberries have become far and away Massachusetts's most valuable crop, surpassing dairy products during the 1980s. They appear to be the one remaining agricultural commodity the state can grow successfully for the modern, global market in foodstuffs.

I suppose we should be proud of this, and I gladly drink more than my share of cranberry juice marketed by Ocean Spray, the growers' coop. But this is not an example of success I can get very excited about because it doesn't suggest much about how to go about reviving a diverse, sustainable regional agriculture. Quite the contrary: it is firmly part of industrial farming. Cranberry bogs are heavy consumers of pesticides and have long been implicated in water pollution, though recent years have seen improvements. I have nothing against Massachusetts boasting an exportable crop, as long as it is grown without damaging our watersheds.[11] Still, it says something about modern agriculture that even in this state, where the cranberry is most notably grown and where cranberry juice is the official state drink, a consumer can't buy a truly ripe cranberry. To get that, you have to buckle on your galoshes and head into the town forest in search of the rare wild bog.

Restoring one small bog and growing cranberries organically might be a good project for each community farm in our region. Since Land's Sake has not tried it, I cannot report on whether or not it is economical to grow and market high-quality, ripe cranberries. It is one more project we would like to try one day.

But I have to admit I would be reluctant to sacrifice the only bog I have ever found in Weston for this experiment—I would prefer to start fresh in some other swamp. That wild cranberry bog is precious to me. There is nothing like going out on a crisp November afternoon with the temperature sinking toward freezing as the sun declines, leaping from tussock to tuft to keep from slumping into the peat and becoming a deeply tanned bog person forevermore. The bright cranberries hang here and there along narrow swales and around the edges of watery depressions in the subdued topography of the bog; they hide down in the cold, wet sphagnum. The low autumn light catches the last fading red maple leaves along the shore, shines along the bare gray maple limbs in the adjacent swamp and glints from the needles of the dark-green pines on higher ground to the west. The cranberry and leatherleaf leaves form a rich carpet of jewel-like red and green among the meadow grasses, which have turned to pale russets and yellows. Who cares that the yield is small and may one day vanish beneath the encroaching brush? For now, the bog is known only to a few in Weston, and the cranberries are reserved for those who have devoted themselves sufficiently to the search. These berries have a fleeting, "ambrosial" flavor that Thoreau remarked in wild fruits long ago, a virtue that would be lost entirely if they were the product of cultivation, no matter how virtuous.

The last time I visited the cranberry bog it was swamped with high water. By dint of heroic leaps my brother and I were able to collect exactly five cranberries, one for each of a disappointed party of pickers who could not venture forth from the firm ground. The cranberry party that day included Brad Botkin's three-year-old son Elijah, who ate both his and his father's berry. That day, five cranberries were plenty.

Blueberries

Half a mile to the northwest of the cranberry bog, through Jericho's stands of red maples, pines, and oaks that were once hayfields, pastures, and scattered woodlots, lies another little swamp. The old cart road curves around it, crossing a culvert over the deep ditch that formerly drained it. This swamp is not as wet as the cranberry bog, and much of it dries out by July. Instead of supporting sphagnum, leatherleaf, and cranberries, it is overgrown with winterberry holly, a few red maples, and a dense stand of highbush blueberries. The blueberry bushes tower eight to ten feet tall and form a nearly impenetrable thicket with narrow, twisting alleys and occasional small, hidden openings where, after tunneling through tangled branches and collecting a head full of leaves and twigs, a person can stand upright to pick berries. In a good year the bushes are loaded with blueberries for several weeks. These berries are not as large as the cultivated varieties of the same species *(Vaccinium corymbosum),* but they are tart and have subtle variations in flavor from one bush to the next. Their colors range from powder blue to nearly black, their shape from squat pumpkin to drooping pear. Even in a poor year, a determined picker who is inured to mosquitoes can fill a quart container—enough for a single pie to mark the season.

Like grapes, blueberries bear on shoots from year-old branches, and so they do better after being pruned to promote new growth. One winter I decided to cut back a few of the wild plants to see if this would improve the yield. I pruned three or four bushes just an hour after dawn on a cold Saturday morning in early March, with the flooded swamp conveniently frozen underfoot. As I was pruning the blueberries, removing a few of the old woody stems from the base of each plant, I discovered squarely trimmed stumps among the whorls of living branches: somebody had done exactly the same job several decades before! I was working not simply in an overgrown hay meadow (as I imagined it might have been), but in what had once been a semicultivated blueberry swamp, as I envisioned it might become. The human history of that place was much closer to the surface than I had thought, residing in the current layer of seemingly wild vegetation. Not so long ago, people in Weston troubled themselves with caring for blueberry swamps. This was news to me, and good news. I like to think well of my predecessors.

In Massachusetts, wild blueberries are ubiquitous and were once a notable crop. In addition to the highbush variety, which grows in the wet places, both the early lowbush *(Vaccinium angustifolium)* and late lowbush blueberries *(V. vacillans)* are found throughout the uplands, now mostly shaded by woods. In his essay "Huckleberries," Thoreau called these plants "the most persevering Native Americans, . . . a sort of miniature forest surviving under the great forest, and reappearing when the latter is cut." In his day, of course, the great forest largely *was* cut, but he observed that nature "compensates us for the loss and refreshes us with fruits such as the forest did not produce." In our day the forest has mercifully recovered, and so in Weston lowbush blueberries are hard to find in concentrated abundance anymore, although the spindly plants still carpet the forest floor, waiting for the next hurricane. They bear millions of berries every summer, of course, but the fruit is spread so thinly across the forest that it requires great zeal to pick even a quart or two. Blueberries require sunlight and fire to thicken and flourish. These days many people pick a casual handful but seldom head to the woods and fields in August to collect in quantity.[12]

It was not always so. Until recently, blueberry picking was a normal part of life in New England. In nineteenth-century Concord, Thoreau described throngs of pickers heading for the huckleberry and blueberry fields every August, on foot and in hay wagons: "Women and children who never visit distant hills and swamps on any other errand are seen making haste thither now with half their domestic utensils in their hands. The wood-chopper goes into the swamp for fuel in the winter; his wife and children for berries in the summer." After his death, Thoreau was eulogized (and scolded) by his friend and mentor Ralph Waldo Emerson as aspiring to nothing higher than "captain of a huckleberry-party." Thoreau might well have replied that no higher calling exists. What is called in this district the huckleberry *(Gaylussacia baccata)* has large, black berries and a coarse flavor that few modern pickers seem to find as palatable as that of blueberries, but the two were apparently picked with equal relish in Thoreau's time.[13]

According to Thoreau, the huckleberry crop was more important than the apple crop in Concord. If this was truly the case it is not reflected in any agricultural statistics I can find, perhaps because comparatively little of the berry crop was marketed—in that way it may have represented one of the last vestiges of the local subsistence economy of colonial times. Yet some part of the huckleberry crop was picked for sale, a practice Thoreau deplored because it turned the fields to commerce. This left the best part of the fruit (which is the health and happiness of picking them yourself) to rot on the bushes, in his view. Be that as it may, I know that blueberries were still being diligently picked by children into the early twentieth century in Weston. Doug Henderson has told me stories of where the picking parties would go in his youth. The Weston historian Alice Fraser wrote of picking blueberries as a girl in the 1910s and selling them for fifteen cents a quart. They were sent from a neighbor's truck farm to Boston Market along with the rest of his produce. Picking berries to earn petty cash was a common experience among children in Weston up to that time, and I trust that while they were at it they got some of the best part of the fruit, too.[14]

Today, the Weston child who has picked enough wild blueberries to fill a single pie is unusual. This is partly because of a cultural change, no doubt, but it is mostly the result of a parallel ecological change. The forest has returned, and the ragged open places are gone. In 1940, Massachusetts reported more than 1,800 acres of commercial wild blueberries, mostly in the hilly western part of the state (Middlesex County reported but 53 acres). By 1992, only 330 acres of wild blueberries were harvested in the state, none of them in Middlesex County. The wild blueberry industry is still important in eastern Maine and on down into Nova Scotia. In Maine, you can see vast wild blueberry monocultures, which are burned every other year to renew growth and eliminate competition. I passed hundreds of acres sparkling with blueberries on a visit there one August but was warned by my host not to touch them because the fields (belonging to his neighbors) had just been sprayed to control worms. I was also informed that pesticide runoff from the blueberry barrens was blamed for polluting rivers and coastal estuaries. It was like looking at hell on earth, to see wild blueberries prostituted and brought to this debauched condition. The evil days that Henry Thoreau foretold are now upon us. Like cranberries and apples before them, wild blueberries have been transformed from a common part of every local landscape to a commodity concentrated in one region of specialized production and heavily sprayed with poisons. Thus they have acquired the defects of all modern crops. *"Sic transit gloria ruris,"* wrote Thoreau. "The wild fruits of the earth disappear before civilization, or only the husks of them are to be found in large markets."[15]

We should not pine for the unkempt landscape where once blueberries flourished for the picking on every vale and hill. As Thoreau pointed out, these berries were poor compensation for poor care of land: the by-product of degraded pastures returning to forest and of woodlots too frequently clear-cut and burned. We don't want that world back in every particular. But there may be other, more benign ways of managing a portion of the landscape to encourage blueberries, along with a variety of other now-neglected forest crops. We should explore the possibilities of various agroforestry combinations, well suited to different parts of the land. Agroforestry systems seek to combine forest plants with other crops. These might include species growing together in several layers from the lowly herbaceous up to an open canopy, or species growing in successional sequences that mimic natural forest disturbance and recovery. There is plenty of precedent in both our Native American and premodern European heritage for handling part of the landscape in this way.

At first, the English tradition does not look very promising. Europeans did not integrate their food production system with the forest nearly so thoroughly as many Native American peoples, although a wide range of foods and other products were certainly once gathered from partially wooded commons in Europe. Once in New England, the white inhabitants never made a serious long-term effort to manage the forest for food, even though they did rely on it heavily during the earliest days of settlement. The European rural tradition, however, does include a technique of woodland management that has potential for agroforestry: coppicing. At its simplest, a coppice is a woods composed of trees that are cut periodically as they resprout from the stump. Several of the most common New England trees, including oaks, white ash, red maple, locust, hickory, and chestnut, sprout very well. During the nineteenth century many "sproutlands" were regularly cut clean on rotations of about thirty years, which is a crude form of coppicing. Is there anything here that could be adapted into more complex agroforestry systems?

Coppicing in Europe is an ancient and subtle art. The technique is thousands of years old in England, where the few coppiced woodlands that remain frequently descend directly from the wildwood that once covered most of the country. Sproutwood, typically hazel but also ash, linden, and many other species, is cut on a short rotation of five to fifteen years. This produces the copse, or spinney, like the one around which Pooh and Piglet once tracked the multiplying Woozles and Wizzles in the snow. The slender coppice wood was used by English villagers to make thatching spars, woven hurdle fencing, and wattle and daub housewalls. It was also bundled for firewood. Surviving coppice woods in Britain are very rare—most of the woodland in that country has unfortunately been converted to plantations of exotic softwoods during this century.

The coppiced underwood was usually grown beneath an overstory of large timber trees called standards, usually oaks. These trees were widely scattered so as not to shade the coppice too severely and were selectively harvested on a much longer rotation—sometimes longer than a century. In this way, tree crops were simultaneously grown on the same land in two layers on two separate rotations, a long one in the canopy and a short one in the understory. The oak standards supplied timber for buildings and ships, while the coppice supplied small wood for a multitude of purposes. The herbaceous layer included such wildflowers as bluebells and primroses, which burst forth in great profusion to carpet the ground when the coppice was cut. Faith and I have visited such woods in May, and it is a sight not soon forgotten. Certain species of wildlife, nightingales, for example, also adapted to coppiced woodlands. In fact, the recent restoration of a handful of traditional woodlands in Britain has been led by wildflower and ornithological societies, much in the way groups of citizen ecologists have sought to restore patches of prairie and savanna in America. No doubt there is a wattle and daub society somewhere in England, but I do not know how much ancient woodland they have rescued.[16]

This multilayered approach to the woods suggests many agroforestry possibilities for New England. I am hardly calling for a wholesale conversion of our forest to coppice, but rather for a few experiments at the fringe where farm and forest meet. The challenge is to find workable combinations of high and low species that provide a range of harvestable products. Nut tree species that also produce high-quality hardwood timber are obvious candidates for our standards, to be grown on a wide spacing so they can spread their crowns and be pruned up to twenty feet for a clear butt log. White oak, shellbark hickory, black walnut, northern pecan, and blight-resistant American chestnuts (should they become available) are all excellent choices. We could plant the best-yielding varieties of these species and eat more nuts. We would presumably need to eat more squirrels, too—if our Disneyfied society can stomach it. We currently kill these creatures profligately with Jeep Cherokees, why not kill them judiciously for meat? I understand they are delicious stewed.

For the coppice layer we might work with black locust for fence posts, white ash and hickory for tool handles and walking sticks, various specialty hardwoods that are in demand for woodcrafts, and a variety of nuts and berries. For the herbaceous layer we might try lowbush blueberries and perhaps medicinal plants such as ginseng and mayapple. Plants like locust, basswood, and blueberry, whose flowers produce uniquely flavored honey, are also good. What if in the highbush blueberry swamp in Jericho we were to replace the now-ubiquitous red maples with black ash, which was once found in these parts but has been extirpated by wetland drainage and overharvesting for basket wood? What about coppicing basket willows, pussy willows, or red osier dogwood in other wet areas? The exploding deer population must be controlled for coppicing to succeed, but again, this is less a drawback than another gustatory opportunity.

We have tried a few such agroforestry experiments at Land's Sake. Our most interesting trial was not with hardwoods but with nonsprouting softwoods. On a piece of private land we thinned a stand of middle-aged white pine to widely scattered, straight stems that would grow out to make superb, clear lumber. Because the crowns of these pines were so high and small they cast very little shade, making them ideal standards. Underneath, we planted rows of Christmas tree seedlings, which grew well in the absence of competition from grass. The dappled shade did not seem to retard their growth. The forest floor was weed-whipped by hand a few times a year, discouraging heavy brush before it could get started and encouraging lowbush blueberries to fill out from the straggling plants that had been growing in the pine duff. This system, invented by our foresters John Potter and Jeff Bopp, was a very promising low-maintenance, three-layer agroforestry combination.

The project went along splendidly for a while, but it came to a bad end. The first few years after a dense stand of pines is thinned, the trees, being shallow rooted, are extremely vulnerable to windthrow. A year or two after we began the project, a hurricane named Bob flattened most of the stand. Because of the high cost of cleaning up the destroyed timber, the landowners who were sponsoring the experiment lost interest after the storm. There it sits to this day, untended. The episode illustrates something other than that thinning white pine is risky. Agroforestry and tree crop experiments take a long time to bear fruit and have always languished on the margins of agricultural and silvicultural research. Many have been started in the past century but few seen through. We need institutions that will make a small investment and carry it on, running a wide range of trials. As in the case of cider apple orchards, community farms could be those institutions. Just avoid long-range projects on private land.

Land's Sake has tried another agroforestry experiment that is turning out better. Once again, the main crop was Christmas trees but planted this time in an old grassy field. A frequent problem with such plantations is that the resurgent grass proves a formidable competitor to the young tree seedlings. Timely cultivation is expensive and often neglected when the payoff is five to ten years down the road. We solved this problem by planting pumpkins and winter squash as annual companions for our trees during their early years. We left just enough room to cross-cultivate with the rototiller between the rows of pumpkins and trees. After the tree seedlings had made their annual growth, the pumpkin and squash vines swarmed all over them but didn't seem to injure them any. The income from the cucurbits more than paid for the cost of cultivating the trees. After the trees were well established and taking up more room, we sowed perennial white clover. We have begun harvesting Christmas trees from this planting and plan to sell most of them as cut-your-own. Our only regret is that we didn't think to replace every fourth or fifth tree with an apple whip when we planted them, so that when the Christmas trees were harvested we would have a cider orchard already beginning to bear. Experimenting with agroforestry is a slow process.

The Native peoples of New England are a much more profound model of how to live with the forest. We can learn a great deal about ecological relations from their ways of living in this part of the world, even if we do not imitate them directly.

Native Americans collected an impressive range of foods from the surrounding environment, moving about within their territories to take advantage of the seasons. But it wasn't simply that they knew where the wild things grew and attuned themselves passively to nature. Native peoples throughout America tended their landscapes in sophisticated ways, and the inhabitants of southern New England were no exception. Whereas in the European tradition a fairly sharp line divides farmland and forest, almost all aspects of indigenous husbandry and foraging were more closely derived from and connected with the forest. Some historians have described such landscapes as domesticated ecosystems, a good description as long as one does not take it to mean that humans were dictating the living arrangements of all other species. What was practiced seems to have been a broad manipulation of ecological zones, in order to encourage the production of a wide range of useful wild plants and animals. New Englanders are not likely to return to this way of gaining their subsistence, and I do not believe they are required to do so in order to be ecologically successful; but the Native system does have something to teach us.

Agriculture is a fairly recent arrival in New England. From about one thousand years ago until the English husbandmen appeared three and a half centuries ago, Native peoples of the coasts and river valleys of southern New England were clearing land and tending crops. Cultivated crops are estimated to have supplied more than half of the caloric intake of Native farmers, but these crops occupied only a small part of the landscape and were integrated into a much larger, diversified subsistence economy. Throughout coastal southern New England, Native women hoed their gardens of corn, beans, and squash, an acre or two for each family. Their planting grounds were located on dry, sandy soils formed in glacial outwash terraces beside rivers and ponds. Groups of horticultural people lived in the Concord and Charles river valleys on either side of what is now Weston, and it may be that on occasion some of their fields reached into the southern part of the present-day town; but we can safely assume that the area's rocky uplands and swamps served mostly as peripheral hunting and foraging territory. The Natives did not farm the stony glacial till soils that dominate Weston, so the hills remained covered in heavy timber. Native women preferred the sandy ground not because the soil was particularly rich, but because it was easier to work with stone and shell tools. These warm soils were well suited to planting long-season corn and squash crops in early May while there was still enough soil moisture to get them started. Once established, the tropical cultivars could withstand dry summer weather.

Native people kept no domesticated livestock and did not rely on manure to maintain the fertility of their fields, as their English successors did. They may have fertilized their crops from the burgeoning runs of alewives that came up the rivers into ponds to spawn in the spring, trapping them in weirs and hauling them to the fields in baskets. The use of fish fertilizer is disputed among anthropologists and historians. I am among those who believe that ancient farmers everywhere were familiar with this simple trick. But whether or not they made use of fish to prolong the life of a field once cleared, the Natives relied primarily on shifting cultivation to restore the fertility of their planting grounds. That is, they borrowed their farmland from the forest for a short time only. After several years of cropping, they allowed the land to return to trees, meanwhile clearing a new field and burning the slash to release the nutrients that had accumulated in the vegetation during several decades of regrowth. After the field was abandoned, they often encouraged crops of strawberries, followed by blackberries and raspberries—like experienced shifting cultivators everywhere, they exercised some control over the successional regrowth. Finally, the cropland was released to pitch pine, which thrives on dry, sandy, frequently burned ground, and so occupied the recovery phase of this cycle of shifting cultivation. Our best guess is that these cycles took something like half a century to go around. In this way, perhaps 20 percent of the native landscape was occupied by a horticultural rotation—but only one-quarter or one-fifth of this acreage was planted at any given time, while the rest recovered in pitch pine. Sandy plains along the rivers south and west of Weston conformed to this description when the English arrived in the 1630s.[17]

Native people lived in the region for thousands of years by foraging before they took up farming, and these older food gathering strategies were not forgotten when farming was adopted. There is evidence that the yields of many of the foraged crops were enhanced by manipulation of the landscape in some way. Probably the most productive zone, from time immemorial, was the wetlands. Native people were unable to cultivate the rich alluvial soils within floodplains because they were too wet, but they did rely on these wetlands and meadows for a large part of their diet nevertheless. Besides the fish, turtles, snakes, mussels, and clams that were taken directly from the ponds and streams, large flocks of migratory waterfowl were hunted when they passed through the marshes in the spring and fall. From meadows along the waterways, Native women gathered arrowhead tubers, cattails, rice, and cranberries, along with reeds to make the mats that covered their long houses. Native people may have set fires to periodically sweep the meadows clear of the less desirable shrubs and trees.[18]

There is little question that the Native people burned some part of the uplands, although again there is debate among scholars about how much. My opinion is that they burned a great deal. I believe that fires were set at varying intervals over a large part of the uplands of coastal southern New England, going back thousands of years. These fires would have been light and frequent across the parts of the landscape underlain by coarse, drought-prone glacial outwash and would have penetrated less frequently into the hillier regions underlain by moist glacial till—but often enough, as we have seen, to effectively keep sugar maple and other fire-sensitive northern species from gaining more than a foothold. The Natives had two closely related reasons to use fire in this way. The first was to maintain a forest that was dominated by white oak, chestnut, and hickory, having an open canopy with wide-spreading crowns. These nuts were an important food source in their own right, and there is evidence throughout the Northeast that Native Americans planted and nurtured nut trees. The resulting "mast forest" also provided rich feed for animals, the second reason for employing fire. The Native people relied on deer and other game for an important part of their protein, and so it was in their interest to create and maintain open areas where animals could browse on the regrowth. Burned places were also good for berry picking. Many open hilltops in the Weston area were named Hirtleberry Hills by the English in the seventeenth century. The women and children who made haste to the blueberry fields in Thoreau's Concord were carrying on a tradition that was thousands of years old, whether they knew it or not. Indeed, late-twentieth-century inhabitants of the region are probably the first New Englanders in the last several hundred generations so clueless about our surroundings that we don't habitually pick blueberries.[19]

Thus, Native Americans modified and made use of the forested landscape in a number of ways, depending upon the varying qualities of the soils, vegetation, and wildlife they had to work with. They promoted nut trees, berries, and game in a complex mosaic reaching from the river meadows to the hilltops. When they took up farming, they cultivated land in rotation with a part of the forest. In a word, they practiced agroforestry, on the scale of the entire landscape. This was a tradition worthy of our respect.

The long-term success of Native peoples at cultivating the landscape for forest and wetland production ought to make us stop and think about whether this might not be a better model for agriculture in New England than European mixed husbandry,particularly given the mixed record of pasture and tillage farming here. I am not suggesting a return to the Native system, which could not support our level of population. I am suggesting that we devote part of our agricultural research effort to the possibilities of improving oaks, hickories, chestnuts, blueberries, and other native species as substantial food crops. One can imagine agroforestry systems that would combine the best of both worlds, native trees along with European grains, grasses, and livestock. Are there arguments in favor of such "natural systems agriculture" for the region?

If farm systems are to last centuries or millennia, they must reproduce many of the features of natural systems—that is one of the central tenets of sustainable agriculture. Some believe that tillage and grazing, the ancient arts of husbandry, have proven themselves too ecologically disruptive to meet this test. Tilled land is highly productive because it is intensively cultivated, which is to say, ecologically simplified. Biological production is concentrated in one species at a time, competition from weeds and pests is strenuously beaten back. The reward is unnaturally high yields of crops, which comprise a few choice parts of a few pampered plants bred to overachieve in a sheltered environment. The storable surplus agriculture creates has caused cities to rise and civilizations to follow. In time, many agricultural societies have fallen at least partly because of the degradation of the croplands that supported them. Ecological oversimplification increases vulnerability to soil erosion and pest invasion, and therefore tilled, monocropped land often invites disaster. Grazing livestock are also ecologically simplifying. Around much of the Mediterranean world, for example, grazing and fire maintain an ancient scrub environment that is considered degraded by many ecologists. The world contains vast stretches of quite stable but rundown and stingy grazing land that was once prime forest, arable land, or both. Environmental degradation has been one of the fundamental engines of world history since the Neolithic.[20]

For those who are convinced that conventional tillage is irredeemable on most soils, the sustainable alternative requires the invention of new agricultural systems that more directly mimic natural ecosystems that have proven themselves well adapted to the world, region by region. In forest country, this logically suggests a mixture dominated by food-bearing trees and shrubs. More than half a century ago, J. Russell Smith's *Tree Crops* argued that in the long run only trees are ecologically appropriate for the steep slopes and high rainfall of many forested regions. A more moderate expression of this idea is the embryonic field of agroforestry. It proposes systems that combine tree crops and agricultural crops in complex, dynamic mixtures. Emphasis is placed on native perennials, often involving several layers of plants in mixtures that may change successionally over time.

These radical ideas have been adapted to the prairie region by Wes Jackson and my other former colleagues at the Land Institute in Salina, Kansas. There, the concept of mimicking the natural prairie has inspired research to develop perennial polyculture grain crops. The world's principal cereals are all derived from annual grasses, adapted to growing in disturbed areas. Their cultivation requires profound ecological disruption. Is it not worth examining whether alternative grains could be bred from perennial grasses that hold the soil? This is an especially important quest, given the overriding economic and cultural importance of grain consumption and the ecological consequences of grain production on many of the world's most productive and most vulnerable grassland soils.[21]

Obviously I believe that these experiments are well worth pursuing. Yet I am not convinced that tree crops and agroforestry are ecologically *necessary* to sustainable agriculture in New England. Much as I love trees, the argument that because the region was naturally forested only woody perennials can hold the soil here strikes me as too wooden an application of outdated ecological dogma. The northeastern United States closely resembles northwest Europe, where mixed husbandry has proven reasonably stable. I believe that the soils and climate of Northeast America can support well-managed and carefully adapted tillage and grass farming over a substantial part of the landscape indefinitely. But there is nothing wrong with including tree crops as part of that mix, if workable methods can be found. The ecological strengths of the forest are well worth examining and imitating. Agroforestry and other forms of natural systems agriculture may be an important part of our future.

When we pick cranberries and blueberries and boil maple syrup, my thoughts return sometimes to the depth of human occupation of this region. Native people lived on this land for millennia before we took it from them, and we ought to remember that. The notions that they passively lived in harmony with nature and took only what nature would give aren't inspiring to me. Such fanciful eco-Indians are noble savages trapped in the wistful, patronizing mold of Rousseau. That image dismisses these people as intelligent human beings fully aware of and able to manipulate their environment, albeit within a spiritual framework that united the natural and supernatural in ways that are lost to our thinking. When I think of them *caring* for the world in this way, by flint and fire reshaping the forests and meadows for thousands of years to be richer in food for humans, the hair on my arms stands on end. These people lived in this land for a long, long time. Our prospects suffer greatly by comparison, given our record to date. We ought to be inspired by their example when we think about how to *use* this land. Instead, we strand them in the fabled wilderness we have lost.

On the east side of Weston is a rocky outcrop overlooking the Charles River valley. It is called Doublet Hill. Here I discovered my bearberries a quarter century ago. From Doublet Hill there is a splendid view over the city of Newton all the way to the Prudential Building and Hancock Tower in downtown Boston ten miles to the east. To preserve this view, the Conservation Commission has Land's Sake clear trees and brush along the south and east sides of the hill, just below the summit. Doublet Hill, no doubt, was once burned by the Natives and was covered with blueberries. In my fantasies, we would instead clear the top of Doublet Hill all the way around and then set fire to that knob every few years to keep it clear in the ancient, time-honored way. I like the thought of people in the city below looking up on a spring evening to see a hilltop in Weston all in flames. (The Environmental Protection Agency would have a fit, of course: automobiles and power plants have so fouled the air that moderate, sensible uses of open combustion are now discouraged.) Having burned, we suburban agrarians would climb Doublet Hill on August evenings to pick blueberries while oblivious urban New Englanders commute in droves on the beltways below. We would relish our berries, anticipate our dessert of pie and wine, contemplate the people who picked berries on this hill when the glacial ice had hardly withdrawn over the horizon, and ponder what kind of people we will have to become in order to be here to welcome the ice when it returns.

Land's Sake forester John Potter stacking wood, 1990. (Susan M. Campbell)

Chapter 5

The Town Forest

Trees, unlike fruits and vegetables, are a tough subject. Most environmentalists readily endorse organic farming. There is broad appeal in growing food in ways that don't pollute air, earth, and water and that rely less on fossil fire and more on everyday sunshine. People relish the idea of healthy food as the diet of a sustainable future. My advocacy of local food systems is hardly a controversial position in the environmental movement: nobody objects to cutting down broccoli in its prime.

Cutting into the forest is another story. When you leave the garden for the woods, you enter dangerous territory, where a preservationist hides behind every tree and a utilitarian lurks under every rock. In the forest, two deeply held environmental imperatives collide: preserving wild ecological integrity on the one hand, and relying on renewable biological resources on the other. This old struggle within the conservation movement keeps resurfacing in various guises, and the conflict is only growing more intractable as our species disturbs the biosphere on an increasing scale. We must find ways to protect our fellow species on this planet and give them room to flourish alongside us. For most environmentalists, protecting biological diversity has been closely linked to, if not identical with, the preservation of wild natural areas, large and small. Mountains, deserts, prairies, estuaries, and coral reefs are all critically important, but *forests* stand at the heart of the matter for many of us. We have a sense that if we lose the trees, we are finished.

At the same time, most environmentalists are attracted to wood. We love wood furniture, musical instruments made of wood, warming ourselves by wood fires. If we don't rely on a renewable resource like the forest for a significant part of our material needs—lumber, paper, and energy—we will be left to rely instead on synthesized products whose environmental impacts are often far worse: concrete, steel, plastic, oil. By reducing consumption and recycling wastes we could go a long way toward mitigating the demands we make on natural systems, and that would be all to the good. Nevertheless, if we face a future in which fossil fuel use must be cut dramatically to avoid bringing the ecological disasters of poisoned air and disrupted climate down on forests—on all forests, wild or managed—it seems inescapable that we must turn to the *local* biological resources immediately surrounding us to meet a larger proportion of our needs. And many of us would rather live in contact with wood than in an alien world made up of artificial, synthetic substances, anyway. We would like to become "ecosystem people," in touch with the natural world at our doorsteps.

So the question is, can we accommodate both of these imperatives, biodiversity and renewable resources, within the same forests? These imperatives cannot very well be separated—each demands a larger part of the world than the other can spare. Both need to be satisfied in every region, at every scale, to make any difference. They must overlap. This means sustainable forestry.

The idea of sustainable agriculture feels comfortable to most environmentalists, but mention sustainable forestry and you arouse deep misgivings. To many it sounds like an oxymoron, by the following line of reasoning: "Agriculture creates a humanized part of the landscape and can be made sustainable only to the limited extent that we learn to incorporate ecological principles into it. We only ask it to stay within decent bounds and to not mess things up too badly. The forest, conversely, is the natural part of the landscape. Any human 'management' would seem to be in fundamental conflict with allowing ecosystems to function freely in ways that have evolved over eons to support the full diversity of indigenous species. Nature is wiser than we are about what should grow in the forest."

Even if the possibility of sustainable forestry is granted in principle, forestry in actual practice has made such a bad name for itself that yoking it with the word *sustainable* is still a stretch. Much of the logging going on around the world today has nothing to do with making use of forests in ways that begin and end with ecological integrity but is straightforward industrial extraction, whether naked or clothed all in green, oh. In many places we see sheer rapacious assault, in others intensive management designed to churn out a handful of commercial species at the expense of the rest of the ecosystem. Most environmentalists can agree to oppose that sort of blatant forest exploitation, but the hard question remains, What is the better alternative? Must it be wilderness for safety's sake, or is some kind of sustainable forestry really achievable? Given this background, what is the most environmentally responsible role for a few thousand acres of forest in a small town in New England?

I began thinking about these issues during the late 1970s when I was living in Concord, Massachusetts, with my friend Nat Marden on the shores of Fairhaven Bay. The old summer house we lived in was surrounded by 150 acres of oak and pine forest that Nat's grandmother had left to the Concord Land Conservation Trust. We winterized the place. In the living room Nat installed an unadorned, boxy sheet-steel Riteway stove, standard issue for the no-nonsense woodburner of the seventies. Nat had a Jonsered chainsaw and a Dodge Power Wagon, and he brought in loads of oak deadwood from the surrounding forest, which we split and stacked on the back porch until we had seven full cords. This fed the hungry Riteway all winter. For my part, I bought a Stanley cookstove, made by the Waterford company in Ireland. Fed with short rounds of hard-seasoned oak, the cooker kept the kitchen end of the house warm. It was a great way to live.

I did not have a chainsaw of my own at the time—I preferred hand tools. A common theme in wood heating literature concerns the neophyte who moves to the country and teaches himself to cut and split wood by hand. He keeps at it doggedly for a few years but can't help noticing that the good old country boys prefer chainsaws and hydraulic splitting machines. And so he finally mechanizes. The essay is about coming of age, about shedding post-card romanticism and coming to terms with rural reality. There is a kind of inevitability about the story; a grinding and slightly wistful sensibility that says, "The machine is here to stay and life is short, so let's get on to the better things." But in some cases, rewarding handwork *is* one of the better things in life. I resist the idea that we must choose, irrevocably and absolutely, between a world of machines and a world of primitive hand tools. I love splitting wood with a six-pound maul. I loved it when I was in my early twenties, I love it now that I am in my forties, and I hope to go on loving it into my sixties and maybe even my eighties. It is efficient and relaxing and good exercise. I have, however, changed my opinion about chainsaws.

Like many environmentalists, I learned to hate chainsaws young. They epitomized everything that was wrong with our approach to nature, brought down to the personal level: the snarling machine, obnoxiously noisy and dangerous, all too easily and thoughtlessly used to destroy noble living things. I wanted to have a woodsman's connection to the forest, one that relied on mastery of hand tools and the healthy limitations that such tools impose. This principle sounded good, and it seemed to hold up perfectly in the case of the splitting maul. I set out to fell trees and buck firewood with a sharp ax and a well-filed crosscut saw and to enjoy my work. After all, these simple tools were sufficient to all but wipe out the forests of North America before chainsaws were even dreamed of—surely they should be adequate for more restrained, selective cutting. I soon reached a conclusion: splitting wood by hand is reasonably fast and efficient, felling trees and cutting wood by hand is unremittingly slow, grueling toil. After watching Nat in action for a year, I conceded that the chainsaw is a good hand tool.

The next winter I bought a Jonsered chainsaw and an old Jeep pickup. From Nat I learned to file my chain so exquisitely sharp that the saw fell through seasoned oak from the strength of its own desire, without any downward pressure from me. I learned to despise the first hint of hesitation in the fluidity of the cut and to stop and touch up the chain with a file. While cutting, I kept the bar from touching the dulling earth at all costs. I discovered that much as I loved growing carrots, pressing apple cider, and boiling maple syrup, none of these gave me quite the satisfaction that dropping trees and cutting them into sixteen-inch lengths with a good, sharp chainsaw did. The Jonsered saw, made in Sweden, was durable and well balanced, light and comfortable to work with, throwing out fat flakes of red oak, creamy maple, or pale ash in a redolent stream as the quicksilver teeth sliced down through the wood in smooth, straight lines. I found that sawing created a focused world that was oddly peaceful and relaxing.

The power of the chainsaw to cut organic tissue imposes two responsibilities on the cutter. The first is toward his own safety. Chainsaws are dangerous. Like most young men I was slow to fully comprehend this and to give up the notion of my invincibility, although I believed I had a healthy respect for the saw. I was no crazy daredevil like some guys I've seen, but when I think of the things I did in those years I now realize I was simply lucky not to get hurt. One day an older hand (perhaps all of thirty) with whom I was working remarked, "You're cutting naked." When I insisted I was being quite careful and knew what I was doing he elaborated, "You just haven't bled yet." After thinking about that and recalling a sixteen-year-old friend of Nat's who cut through his kneecap one afternoon, I got some basic safety gear: a helmet with eye and ear protection (hearing loss is cumulative), ballistic nylon cutting chaps, and steel-toed boots. These do not make me invincible, but they do give me some grounds for feeling at ease with a chainsaw.

The second responsibility of the cutter is toward the trees. A person with a chainsaw can do the forest damage in a very short time. When I began cutting firewood it was the era of the energy crisis. Heating with wood was all the rage, and the price of firewood was high. Guys who had no idea what they were doing were running around the woods dropping whatever was easiest to slice up and sell. If the high price of oil became a permanent fixture, as we all believed it would, then it seemed likely that the normal pattern of American market relations to natural resources would repeat itself in the New England forest. When the resource has little value it is largely neglected, and when the resource rises in value it is quickly squandered. Then it is either abandoned again or expropriated by those with capital. Experts were confidently predicting that after the initial blitz, scientific management would take over the woods, and we would see uniform blocks of fast-growing hybrid poplars owned by large corporations efficiently harvesting their captive biomass with whole-tree chippers.

I wasn't looking forward to any of that. There had to be a way to have real, healthy forests and also enjoy forest products. The Land Trust was allowing us to cut up only deadwood, which doubtless seemed harmless and unobtrusive but was just temporizing. It did not address the problem of how to live with the forest on a long-term basis, producing firewood (let alone timber) in an ecologically sound manner. The dead trees would soon be gone, and cutting only what was dead was ecologically questionable anyway: surely a normal, healthy forest needs standing and fallen deadwood to provide habitat for many of its creatures. Good forest management couldn't mean just cleaning up deadwood; somehow it had to mean cutting living trees. The question was, Which trees to cut?

I began reading everything I could find on forest ecology and forest management, ranging from preservationist manifestos to silvicultural textbooks. I soon discovered there was a divergence of opinion about what good forest management meant, or whether such a thing could even exist. I wasn't happy with the guiding assumptions of the technical forestry books: the economic dogma that oaks and pines became "overmature" and should be harvested as soon as their growth rate tails off at about eighty years old; the concentration on a few commercial species; the narrow focus on board feet. There was, however, good practical information in these books about the way forests grow, how to judge the quality of a site, and ways to measure how thickly it is stocked with trees. The preservationists, on the other side, saw the forest as an ecological whole that had intrinsic value, which was good. But beyond a vague admonition to "live lightly on the earth," they didn't often address the question of where our resources would come from while we surrounded ourselves with unspoiled wilderness.

My own conclusion was that forests didn't need to be managed to keep them healthy, that it was doubtful that management could really improve the forest in a broad ecological sense, but that it might be possible to find ways to harvest timber and firewood sustainably without doing appreciable damage. Was there some way to combine a reasonable level of production with ecological restraint? Common ground seemed to lie not between these two camps but off in another dimension, beyond a range of forbidding ideological mountains and into a sylvan valley, under a whole new set of assumptions about our place in the world.

In the short run, at least, there appeared to be a way out of this bind. Many of the forests in Concord and Weston were not mature, let alone overmature—they were middle-aged. Harvesting of timber was not yet an issue. The idea of periodically thinning these woods to produce firewood might be less controversial: most of the trees in a crowded young stand were going to be squeezed out by their rivals anyway as they grew up. Thinning complemented the natural process. The tougher decisions about when and how and even whether the mature trees should finally be harvested could be deferred. Should they eventually be cut selectively to produce mixed-aged stands, or clear-cut in small patches in a rotational manner to produce a mixture of even-aged stands, or simply allowed to remain standing as stately old growth? These questions did not need to be answered immediately, or even answered the same way in all cases. Meanwhile, we had plenty of work before us and a strong firewood market to make it pay. Here was ecologically defensible temporizing that could go on for decades.

So, with my new chainsaw and my old pickup truck, I set about learning the craft of thinning woodlots. Often I supplied homeowners with a few cords of firewood from woods on their own property, cutting out the overcrowded and damaged trees. I also began selling wood to customers in Weston and neighboring towns. It was pickup truck logging, which works well enough in accessible stands. By this time I was back in school and living in Weston again. I would bicycle home from classes at Brandeis, hop in my truck, cut up and deliver half a cord of firewood and be home for supper and my evening studies.

While I was learning how to cut wood on private land in 1979 and 1980, at the height of the Iranian oil embargo, the Weston Conservation Commission had been coming under pressure to put some of its forested acreage to use providing firewood to the citizens of the town. The commission was understandably reluctant to allow chainsaws into the woods. A few of the commissioners were ardent preservationists who believed Weston had absolutely no right to reduce any of its fine forest to cordwood and board feet, ever. Most of the others on the board were not utterly opposed to forestry, but they were wary of the eco-wrath that they thought would surely fall on their heads when the first tree fell. Yet the pressure kept building from townspeople who complained that all that expensive forest was going to waste, so the commission decided to give fuelwood harvesting a small trial. They had the New England Forestry Foundation mark about five acres of firewood at the back of a red maple swamp in Jericho Forest. It was unlikely that anybody's favorite tree was lurking down there. Land's Sake had just been formed, and we convinced the commission that a fledgling local group dedicated to ecological management was tailor-made for this job (which it was) and that I had the skills and experience to handle it (which I didn't). This was about the same time that we also convinced Harvard University that we could grow strawberries on the Case land. We were off and running. In retrospect, it is a good thing that we started small.

I began cutting one afternoon in early spring just as the swamp maple buds were reddening overhead. I had a job ahead of me. There was no way I could drive my pickup into that soft ground—it was accessible by truck only at one corner. I was going to have to set up my landing there and skid the trees out to it. (In logging terminology, a *landing* is a place where logs are piled up to await trucking, and *to skid* a log means to drag it—the terms do not necessarily imply heavy equipment.) I consulted my friends Russell Barnes and Laura Perry, a forester and a mule skinner (the perfect marriage) in neighboring Lincoln. They sent me Hercules, the mule.

Hercules was the largest mule in New England by reputation, and I don't doubt it. Mules are unusual in New England to begin with, but Hercules was a big mule anywhere, and he loved pulling wood—as much as a mule loves anything. Hercules and Company, as Russell and Laura called their outfit, agreed to skid our wood for some ridiculously low fee—I think they just wanted to see us make a go of it. So through the spring I went down to the swamp and cut maple, along with some oak on the upland fringes. When I had a day's work for him (about five cords) Hercules came over from Lincoln and pulled the trees to the landing. By the time the leaves came out and it was time to start farming, the job was done. I told one of my professors about the wonderful creature I was working with, this mule. His eyes grew wide. "So what do you feed him," he inquired, "meat? Isn't that expensive?" Hercules loved that one when I told him. It's always good to see a big mule grin.

The results of mule logging were great to see: all the wood was on the landing, and there was nothing but a narrow trail winding out of the swamp to show where it had come from. If you followed the trail back to its sources, it gradually forked in every direction and arrived at low piles of red and silver slash surrounding weeping white stumps, cut neatly about an inch above the ground. It was a very light thinning, so most of the trees were still standing to greet the spring. On that job I learned how to hinge a tree properly with a steep, shallow, "open-face" kerf, how to choose the direction of the fall so as to make the skidding easy for the teamster, how to use plastic felling wedges, how to limb a tree efficiently—how to handle a saw in a more or less professional manner. On skidding days I bucked the trees into firewood lengths as they arrived on the landing—"slicing and dicing," Russell called it, keeping the area clear for Herc to drag more logs in. I left the bucked logs in a great heap on the landing in May and turned my attention to breaking ground at the Case Farm. In the fall, we hired a few high school kids, split the wood, and delivered it to a list of customers solicited by the Conservation Commission. We learned that wood doesn't season adequately unless split and stacked so the air can get to it, and it made us unhappy to sell improperly seasoned wood to our friends and neighbors. Land's Sake made plenty of mistakes and lost a little money on that job, but we were in business at last, cutting wood in the town forest.

Over the next few years we went on to thin fifty to a hundred cords of firewood from the town forest every winter, a small but respectable operation. Cutting cordwood required no more equipment than a chainsaw and a pickup. We learned to choose our sites. Many of the upland oak and birch lots we worked in the Highland Forest behind Doug Henderson's house were so open and accessible that we went back to pickup logging. They had been pastures and hayfields a century before, after all, and all the rocks had been thoughtfully hauled out of the way for us. In rougher lots we hired Hercules & Co. to do our skidding again. After Russell and Laura moved to

Vermont and took the mule with them, we tried out a small radio-controlled winch that Russell was promoting. We weren't making big money selling firewood, but we covered our costs, including labor, which was good enough.

The Conservation Commission and other townspeople grew more comfortable with the sound of chainsaws in the forest. From time to time people would stumble upon our logging and get alarmed, and we got our share of outraged letters and phone calls. These were the people whose support we needed most, so we took them seriously. I asked the commission to direct these complaints to me. Often I would ask the offended party to meet me in the woods on a Saturday morning when I was out there splitting with the kids—this usually got things off on the right foot. Youngsters happily employed with hand tools strike a better note in the woods than snarling chainsaws. We would go for a walk, during which I listened to their concerns and explained our cutting philosophy tree by tree, exchanging views about the history and prospects of the forest and what it meant to the town. Sometimes people showed me a tree they were particularly fond of that had been marked for annihilation with a bright slash of yellow paint, so I promptly unmarked it. There is plenty of room for compromise in community forestry—the woods are full of trees. By the end of one walk I had an initially skeptical tree lover quizzing me about why certain imperfect trees had *not* been cut. Not everybody agreed with our approach to caring for the forest, but by taking the time to explain what we were doing we made loyal Land's Sake supporters out of people who could have become our fiercest critics.

At home in a small town, we took the attitude that it was good that such people saw themselves as the guardians of Weston's forest. The kids were taught that they should be polite, not defensive, when someone out walking the dog discovered our operation for the first time and demanded to know what the hell we thought we were doing cutting their trees! It was important to remember that although we had a close working bond with the forest that made us feel special, we did not own it any more or less than they did. We worked the woods under the watchful eye of those who collectively owned it and cared about it. I think it helped that Land's Sake was best known in Weston for running an organic farm and for working with kids: it gave our claims to ecological and civic virtue credibility. Again, people are more comfortable with farming than logging, and that was part of our image. I don't believe that good farmers always make good stewards of the forest, but I do know that people were ready to trust an organization like Land's Sake, which refused to spray pesticides on strawberries, to see the forest as more than so much sawtimber and pulp.

But sawtimber was *among* the things we saw when we looked at Weston's forest. Removing "defective" trees for firewood was nothing more than a way of concentrating the growth of the stand into those trees that would produce the best timber later on. The issue of timber harvesting was still out there waiting if we wanted to push it, and we did. Someday, if we really wanted to have a responsible forest program that utilized local renewable resources on a significant scale, we were going to have to start harvesting some bigger trees for lumber. After cutting firewood for a few years, Land's Sake felt ready to move up to selective timber harvesting in some of the more mature oak and pine stands in Weston.

There was a lovely stand of large, dense white pine on the Weston College land that seemed perfect for what we had in mind. It was the type of stand that would yield good-sized timber but remain a grove of magnificent trees after the cutting. Led by Julie Hyde, the Conservation Commission approved a plan to use the proceeds from a timber sale to expand our apple orchard on the other side of the College Pond, planting new trees. The plan both illustrated how good management could improve town forest lands and symbolically replaced tree for tree, apples for pines. The commission had a private forester mark a pine harvest on ten acres. To be on the safe side, they held a well-publicized hearing so that town residents could state their concerns before the first drop of sap was spilled. The commission was not about to push through a timber harvest if there was strong sentiment against it. To our surprise, no one turned up to object.

My main concern about the timber harvest was whether Land's Sake, having proposed it, was up to the task of executing it. These were big trees, and we didn't have a skidder. Then again, Land's Sake had a history of tackling things it wasn't quite ready for. Looking up at those towering pines, I wondered whether history would repeat itself: beyond a certain point, on-the-job training gets a little expensive. So again I called Russell Barnes and asked him to come have a look at the job the next time he was down from Vermont to see if he had any ideas about how Land's Sake could do it. In my experience, Russell usually did.

We strolled the college land on a chilly evening in early spring, just at dusk. I expected Russell to tell me all about some slick new piece of logging technology he happened to be importing from Scandinavia that would be perfect for this job—some little robot gizmo of incredible strength that rolled through the woods with a pine log two feet thick strapped under its belly, leaving no trace. Instead, he walked among the tall, stately pines that had been marked, looking up, shaking his head, getting more and more perturbed. Finally, he turned to me and said, "Brian, if you cut this, they'll hand your ass to you."

My gut tightened. "Why?" I asked.

"You just can't cut big trees like this in a suburban town. It'll be an awful mess. You know how *big* these stumps will look? People will go ape when they see this. They'll never let you fire up a saw in this town again."

"But no one even came to the hearing we held!"

"They never do. They'll come to the hearing you'll have to hold after you do this."

I was prepared to hear Russell advise me that the job was too big for Land's Sake, but not that it was too big for Weston. Russell had been through his own suburban logging wars in Lincoln, though, and I knew he was speaking from long, hard experience. "So what do we do now?" I asked.

"You can't let them cut this," Russell said. "Not this way. What you need to do here is take out all the crooked stuff and some of the smaller sawtimber. Do a real clean, beautiful job. Don't let a commercial logger in here. Get it down to a smaller job that you can handle with the radio winch. Maybe we can find someone with a log truck who comes down to Boston with firewood who could backhaul all the junk pine and tops to a pulp mill upcountry. Then you could at least cover your costs on that and make a buck or two on a few loads of sawtimber."

We never cut the college pine stand. It had been marked the only way the forester thought it would be possible to get a professional logger to bid on the job. These are the kinds of compromises foresters make in the real world of commercial timber cutting. They have to include enough high-grade sawtimber to make it worth someone's while to cut and still leave a stand with plenty of good trees to keep growing. As for getting all the crooked pulp-grade pines thinned out, there just wasn't a market for that in our part of the world. We soon discovered this for ourselves when we tried to find someone who would do the trucking if we were to cut the pine the way Russell recommended. The nearest pulp mills were in New Hampshire and Maine. The best offer we could get was from one guy who said he would pick up our pulp sticks for nothing—the rest of the firewood dealers I talked to all wanted to charge *us* to haul it away. They were used to working in the suburbs, too.

We had reached an impasse. Our management plans for the town forest lacked an adequate conceptual framework beyond firewood thinning, and we lacked practical means to carry them out. The Conservation Commission and Land's Sake put the timber harvesting idea on the back burner for a few years and went back to cutting firewood. Getting into serious logging in Weston without losing our shirts, our asses, or both was going to take some more cogitation.

The Fall and Rise of the New England Forest

How does a group of committed environmentalists such as Land's Sake come to promote timber harvesting as ecologically defensible, let alone desirable? It is no surprise that many people who consider themselves environmentalists react viscerally to the idea of logging in forests that were acquired explicitly for conservation—especially to the idea of cutting mature trees that might otherwise live on for decades or even centuries, culminating in old-growth forest. The cutting of big trees seems to flatly violate the sanctity of the natural forest, to be just another version of the utilitarian hubris which holds that we somehow know better than nature how forests ought to be managed. Environmentalists have fought for more than a century against the abuse visited upon the nation's forests either in the name of or under the guise of such "wise use." There is ample justification for this attitude: the history of Americans' relation to their forest up to and including the present is not inspiring, and the future looks equally bleak. On most public and private lands that are cut, in New England as elsewhere, the industrial model of ecological simplification and large-scale extraction prevails.

Many environmentalists have concluded that the proper approach to New England's second-growth forest is to leave it as much as possible untouched by human hands; to let nature gradually restore the diversity and stability of ancient forest to the region. This policy has two serious shortcomings, however. The first is that it is often based on a misconception of what natural New England forest was and is. The second is that it makes no provision for how New Englanders are to take effective responsibility for living in this part of the world. By taking another look at ecological history we may be able to get a better sense of how today's forest came to be and of the possibilities of sustainable human interaction with that forest.

No one is exactly sure what the original forest that greeted the Puritan settlers of southern New England in the seventeenth century looked like. To my knowledge, only bits and pieces of it have so far been seriously studied, and the overall picture is still very sketchy. In fact, the idea that there was an *original* forest to greet the Puritans is misleading in itself—as is the term *second growth* for what greets those born today. Both terms imply that if we go back to, say, 1600 or so, we have effectively reached the beginning, the first growth, an ancient wilderness that had stood virtually unchanged for thousands of years until the arrival of the Europeans. The popular image is of a great climax forest, tall trees stretching dense and unbroken to the horizon, closely adapted to the regional environment and coevolved with a thousand smaller subdominant and herbaceous species in a self-replicating natural order. In this beguiling fable, Native Americans dwelt unobtrusively, almost invisibly, in the great forest's shadow. They are pictured either as submerged in nature because of their small numbers and limited technology or living in harmony with nature because of the ecological restraint embodied in their spiritual beliefs. The forest was scarcely disturbed by them.

The environmental image of the invading Europeans is altogether different. They are remembered as relentless destroyers of forest because they were farming people who didn't understand the nature of the New World's forest and had no spiritual affinity for it and because they were greedy capitalists. The wilderness was hated and feared by the pious Pilgrims, inexorably logged off by the Yankee farmers and lumberjacks who descended from them, and at last, only belatedly celebrated and defended by a struggling band of romantics and environmentalists. Or so we have believed.

Upon closer inspection no part of this story seems to be quite true. That original, unchanging wilderness never existed. The last pristine natural order to cover New England was ice a mile deep, and that vanished 14,000 years ago. Since then it's been a regular organic riot out there, and people have been in the thick of it from the beginning. This is a wonderful story, one worth telling, but it defies a simple plot of ancient natural stability pitted against modern human disturbance. The natural forest was much more disturbed and changeable in its composition than previously believed. On top of that, for thousands of years Native Americans had a significant influence on the forest that needs to be better understood and evaluated. The English farmers who settled New England towns were not simply destroyers—they also developed an intimate relationship with the forest that should inspire us. In the end, though, they surely failed to conserve the forest in the face of other economic demands on the land. The forest has recovered in the past century, but it is not the same forest as before. This forest of ours is not the second but, as it were, the one-hundred-and-second growth. This, as I will detail in a moment, is the complex and ambiguous natural history with which we have to live.

Our task, then, is not to try to restore the forest to its original condition—that is impossible. Our task is to come to grips with our part in this ongoing drama of natural and human interaction, to determine where our ecological responsibilities lie and how we can finally become culturally mature enough to face up to them, to learn from our predecessors and do better if we can. The point of studying the history of the forest is not to rediscover the state of nature we have lost—it is to come to a better understanding of the world that has been created by nature and people before us. It is to develop a deeper affinity and aptitude for the land we inhabit.

We environmentalists like to think long term. In that sense, the diverse, magnificent hardwood forest that people in eastern North America enjoy today is the exception, not the rule. We now know (or at least strongly suspect) that we live in one of the comparatively brief warm interglacial periods that have regularly interrupted the normally cold 2.5-million-year-old glacial epoch we call the Pleistocene. These glacial cycles are apparently tied to small perturbations in the earth's orbit and tilt, amplified by geophysical systems such as oceanic and atmospheric circulation and carbon dioxide levels, in ways that are not yet fully delineated. For about the past 900,000 years, the warm interglacial periods have come around once every 100,000 years and have seldom lasted much longer than 10,000 years—about the length of time we have already enjoyed this one. After these brief thaws, the earth returns to the icebox. We naturally tend to view our own era, the 10,000-year-old Holocene, as postglacial, rather than as just a brief part of an ongoing cycle. But our moment on earth is probably preglacial as well as post-. By any sensible reckoning we still live in the Pleistocene, and the ice will eventually come rolling back down, reestablishing more typical glacial conditions and obliterating New England's forests once again.[1]

Whenever the glacial ice abruptly melts away, as it did 14,000 years ago, all hell breaks loose behind it, ecologically speaking. Vast territories are opened for colonization, while many species that had been living comfortably in the cool world south of the glacier find their homes becoming too warm for their survival; everything shifts hundreds, even thousands of miles to the north. But ecosystems cannot move intact. The species that make up ecosystems move by expanding their ranges on the favorable edge, sending forth their seeds by different means and at different rates. Although some species are closely coevolved to live with certain others, the evidence is that on the whole ecosystems do not hang together on the trail. Instead, migrating species form new ecosystems as they go.

We have a spotty record of what happened at the end of the last glaciation from pollen cores, extracted from lakes and bogs by paleoecologists. Boreal species like spruce and jack pine that had dominated the full-glacial forests south of the ice front all the way to the gulf coast abruptly turned and fled to the northern half of the continent, mixing with all sorts of different trees along the way. Oaks went from living along the southern fringe of the great coniferous forest of glacial times to becoming the dominant trees in many different deciduous and mixed forest ecosystems throughout eastern

North America, including southern New England. Chestnuts moved slowly up the Appalachian chain to finally join oaks as dominant trees here only a few thousand years ago, only to fall victim to the chestnut blight at the beginning of the present century. They will doubtless recover someday, but we will not see it. Sugar maples and white pines are now common through northern and mountainous parts of the eastern deciduous forest and frequently grow together. But during the past glacial advance these species were apparently confined to small refugia a thousand miles apart; sugar maples in the lower Mississippi valley and white pines perhaps somewhere out on the exposed continental shelf off the Carolinas. The modern communities they have formed are real but temporary. Many other species are similarly promiscuous in their social habits. As species expand and contract their ranges in response to climate change, new ecosystems are created in which they overlap and sort themselves out within a given landscape and disturbance regime. This is apparently not a process that often arrives at a settled conclusion, or climax.[2]

Since the ice last retreated, the landscape has been in a state of flux. Many paleoecologists believe that conditions have never stayed constant long enough for anything like stable, self-reproducing climax communities to develop in the New England region. Climate changes too quickly, new species arrive on the scene, or some other disturbance causes the formation of new ecological communities. In southern New England, windstorms and fires periodically create fresh opportunities for regeneration. As a result, even during periods of relative climatic stability there has been a lot of cycling about in the landscape. The result is a patchy forest made up of tracts of varying ages and species—a shifting ecological mosaic. In our place and time, nature has not produced unique, stable, tightly coevolved communities of species, each perfectly adapted to a certain part of the landscape and replicating itself indefinitely through many generations. Instead, nature supplies redundant materials to put together any number of possible functioning communities on each part of the landscape and depends on small differences in adaptation and on chance to determine which ones will coalesce after each new roll of the disturbance dice. Ecosystems are ephemeral: that is the great lesson of paleoecology.[3]

But natural forces were not the only factor that determined how the forest adapted and readapted to our land over the past 10,000 years. There were also the Native peoples, our predecessors in this part of the world. The idea of untouched wilderness awaiting the Europeans in the New World is particularly inappropriate and misleading in New England because during the present interglacial period there have been human beings living here as long as most tree species, if not a little longer. The hunters arrived on the heels of the retreating glacier, while the world was still tundra, open spruce parks, and meltwater lakes. Dirt-insulated icebergs sat in kettle-holes, slowly wasting away to leave behind bogs and ponds. In Concord, Henry Thoreau recounted an Indian legend of a Walden mountain that once stood as high as the pond is deep. This apparent cultural memory of a very real mountain of ice always gives me goose bumps. It reminds me that people may have walked this land while the very soils were being laid down around them by rivers roaring out of ice canyons in the melting glacier—and considered it their common world.

Researchers are not sure how dramatically the various Native cultures that followed these first people affected the landscapes they inhabited, but it was probably more than previously thought possible by our own arrogant race. Paleoindian hunters may have been primarily responsible for the continent-wide extinction of many large Pleistocene mammals such as giant sloths, giant beavers, and woolly mammoths—the matter is disputed among anthropologists, to be sure, but is quite plausible. Succeeding Native American cultures hunted the surviving smaller forest game as the trees returned and were probably among the leading predators of many animals such as deer for thousands of years. The restoring of wolves has become a touchstone among many environmentalists, a symbol of the wild. It is a fine idea. But when anyone talks about the way wolves formerly controlled deer populations in natural ecosystems without even mentioning human hunters, they are lost in a Pleistocene dreamworld that has not existed for so many thousands of deer generations that nature has, effectively, no working memory of it to restore. New England Indians may also have manipulated the environment to *increase* the size of the herd they hunted, by periodically burning the land. I dare say humans have been the decisive influence in many wildlife populations in New England ecosystems as long as there have been New England ecosystems, counting from the last glaciation.

The New World that welcomed the Puritans was not a densely forested wilderness, but a diverse countryside that had until very recently been populated by farmers and hunters. During the early seventeenth century, most of these people were dying from terrible epidemics of influenza, measles, smallpox, and other diseases that had crossed the Atlantic with the European vanguard, an episode in one of the greatest ecological disasters in human history. Their abandoned, unkempt fields may well have been choked with thickets of brush and young trees. This world may have appeared an uncultivated waste and a hideous spiritual wilderness to English eyes, but we need no longer accept that judgment. North America was not a wilderness—it was inhabited land. Celebrating the vanquished wilderness instead of despising it doesn't get us much closer to understanding the real nature of the place we now know as New England.

The forest in Weston today differs markedly from the native forest that immediately preceded our culture's arrival, but that earlier forest itself was also partly a human artifact, a product of another, much longer history of intervention. It is quite possible that had it not been for Native people and their fires, Weston's forest would have become dominated by such northern species as sugar maple, beech, and hemlock thousands of years ago. On the town's sandiest soils, Native fires appear to have imposed a pitch pine forest that, in the absence of that brand of human care, has since withdrawn southeast toward Cape Cod. We are a long way from fully understanding the way our forest evolved and changed before we arrived, but we are now at least reasonably sure it was no undisturbed, stable climax.

Where does all this recently discovered natural and naturo-cultural instability leave environmentalists, who have long sought to draw a code of human ecological conduct, a so-called land ethic, from an understanding and appreciation of nature's harmony and balance? The old standard of wilderness as an immemorially ancient, ecologically stable community no longer seems tenable, which does put all human disturbance in a more uncertain light. Human disturbance is as old as the New England forest to which we look for our standard. But it does not follow that because nature is disorderly anyway and because the Indians were also active manipulators, modern humans may do with nature as they please, with impunity. It is certainly not true that nature is such a chaotic mess that we are somehow obliged to manage it on a grand scale, by virtue of the matching complexity of our consciousness and our computers. The species we need to manage most is still ourselves.[4]

There is a larger natural order that we would be wise to try to understand and get along with, if we hope to remain an ecologically successful species. But it is an order made up of complexity and change as much as of stability. We don't need to worry too much about preserving forever or restoring exactly any specific natural ecosystems—it probably can't even be done. But we do need to worry (as Aldo Leopold suggested in *A Sand County Almanac*) about protecting all the species that supply the cogs and wheels of these timepieces, and the only way we can do that is to protect the many places where they live now. Beyond that, it is our responsibility to give them a landscape through which they will be able to continue to move and to form new communities as the centuries roll on. As I will argue, this means that we need to protect *more,* not less, forested land than we would if ecosystems were nice and stable. We are by no means absolved of our duty to pay close attention to and protect natural systems and processes, or of our duty to remain humble and cautious in the face of natural complexity. I think we do have more latitude in how we assess ecological degradation and in how much we allow ourselves to tinker than many environmentalists have been willing to grant. Deviation from an undisturbed natural norm is not, in and of itself, inevitably ecological degradation. We need to search our history carefully to discover what kinds of degradation have occurred here, to find the real defects in our culture rather than condemn our mere presence, and to seek remedies.

So, to what degree and in what particulars was the forest abused by our forefathers? The first stereotype of New England's forest is that it was originally a deep, dark unchanging climax of tall trees. The second and parallel stereotype is that from the moment they arrived here the white settlers laid waste to this forest with a deadly combination of religious loathing and capitalistic zeal, in an unrelenting slaughter of forest giants, leaving us only a sorry remnant of that ancient forest. There is truth in this. Our culture has never developed an adequate ethos of conservation, and at one point in our history we nearly eradicated the forest. We have an extensive forest again in New England today mostly by accident and neglect. But if it is true that in many ways our ancestors harmed the forest, they also depended on it. Next to the image in our minds of the cold forest destroyers, we hold a warmer image of those same Yankees mastering an intimate vernacular knowledge of the native trees and how to use them. This closeness is expressed in timber-framed houses and handcrafted Shaker furniture. Indeed, colonial Americans relied so closely on the forest that the period reaching into the mid–nineteenth century has been called America's wooden age. At what point did this forest dependence become forest exploitation? Why were these people so foolish as to destroy something of such importance to them, if indeed they did? We need to understand more precisely what their relation to the forest was and exactly where it went wrong.

Wood was immediately recognized as a natural resource of prime importance by the English Puritans who settled around Massachusetts Bay in the 1630s. As towns were established, wood became a key part of their internal economies, and in some cases an export commodity as well. These people came from a country in which woodland had been reduced to only 10 percent of the landscape by the Middle Ages and may not have covered much more than that since Roman times a thousand years earlier. By the time New England was being settled in the seventeenth century, old England had long been importing its sawn lumber and ship masts from the Baltic and was already beginning the fateful switch from wood fuel to coal that would drive the industrial revolution. In New England these immigrants encountered a much more pervasive forest—it may not have been the uninterrupted, trackless wilderness of our imaginings, but it was still like nothing English villagers had seen for thousands of years.[5]

This forest may have dismayed the Puritans' clerics and deceived their cattle, but it did provide resources without which the new arrivals would have quickly perished. New England was much colder than Britain, especially as the worst decades of the Little Ice Age set in. In the terrible winter of 1641/2, Boston Harbor froze to the horizon. Plentiful fuel was essential for survival. Early New Englanders burned staggering amounts of wood in enormous fireplaces and still shivered, until they perfected new saws and axes to cut large trees into small bits and learned to heat small rooms with more efficient shallow fireplaces surrounding central chimneys. Fuel was not all the forest supplied. Imported English goods were expensive, especially as New England farmers lacked a staple export crop like tobacco to sell in exchange. The settlers manufactured everything they could from the materials at hand. For these reasons New England's domestic economy and even its early industrial economy revolved much more around wood than England's did. I think we can rest assured that the average Puritan farmer soon came to regard the New England forest gratefully as another sign of the hand of divine Providence in providing for his chosen people, rather than as a hateful, howling wilderness. The only really dangerous elements in the native forest were the expropriated, decimated, increasingly desperate Native people.[6]

Whatever we think of the newcomers' dealings with the Natives who afflicted them in their New Canaan, we can certainly admire how quickly New England townsmen gained a comfortable understanding of the surrounding forest and learned to make use of its wonderful variety of trees. Most of the necessary skills and tools were perfected within a few generations. The tree of choice for many purposes was white oak, which worked a lot like English oak and which fortuitously was among the most common trees in the new land. If the Puritans had known that white oak's prevalence was largely the result of centuries of Indian burning, they would doubtless have seen the divine hand of Providence in that, too, working as always in their favor. Oak was the prime framing material for buildings and bridges and the preferred wood for chests and other furniture. It is instructive that New England joiners soon switched to using *riven* boards for their work, in contrast to the sawn oak boards from central Europe that were almost universally used in England. This tells us that straight-grained trees that had grown in forests with closed canopies were available in many towns, as well as the knottier, wide-crowned specimens that grew in the more open areas that had been frequently burned by the Indians. Such open-grown oaks may have contained nicely curved timbers for shipbuilding but would have been difficult to rive into straight boards for cabinetry. Oak was also first-rate firewood, often cut from the tops of trees felled for their timber, since branches were easier than massive trunks to work up into cordwood length with hand cutting tools.[7]

Besides its domestic importance, oak was in demand for export, which created conflicts over the use of early town common forests. White oak staves were shaped by coopers to make watertight barrels (or pipes) for use both at home and for the flourishing wine and rum trades in other corners of the Atlantic such as Madeira and Barbados. Sugar could be shipped in more porous red oak barrels. By the late seventeenth century, New England merchants were playing a leading role in the expanding Atlantic economy that revolved mainly around sugar production in the West Indies. New England ships were themselves made primarily of oak. Even in inland towns like Weston, many farmers earned a bit of cash by going to their woodlots in winter to cut pipe staves and ship timber. But the great bulk of their wood and timber was consumed by the farm households themselves or by the dense networks of local artisans and tradesmen.[8]

New England houses were commonly sheathed with either oak or pine boards and floored and paneled inside mostly with pine. Every town soon had several small sawmills that supplied boards for local use—I stumble on a long-forgotten mill in almost every neighborhood I research. White pine was the preferred tree for sawn lumber from an early date, but it appears to have been fairly unusual around Massachusetts Bay, scattered on hummocks in the swamps. The big stands of tall pines that supplied lumber to the Caribbean sugar plantations and mainmasts to the Royal Navy's West Indian fleet lay beyond the Merrimack River, in New Hampshire and Maine. Pitch pine was the more common pine of the dry, frequently burned sandy glacial outwash plains of southern New England. It yielded pitch, tar, and turpentine as well as boards and made a hot fuel worth about three-fourths the value of oak.

Wood objects and wood artisans were ubiquitous in the colonial economy. There were wainwrights, housewrights, joiners, coopers, and clockmakers, to name a few. Every tree species had some specialized use—even lowly willow yielded charcoal that was preferred for making gunpowder. Abundant, rot-resistant chestnut provided a lighter substitute for oak in buildings and fences—several surviving colonial houses in Weston were framed with this now-lamented tree. Either springy white ash or tough hickory was chosen for handles, depending on the tool. The iron froe that split clapboards, shingles, and fence pales was driven by a maul of indestructible black gum, or beetle-bung, so named because it was also used to drive bungs into barrels. There is good evidence that black ash was once found in swamps in the Weston area. The English learned of its excellence for basketmaking from the Native Americans. Dense white cedar swamps also appear in the early records of many towns. Cedar was prized for providing watertight shingles and clapboards, along with moth-proof chests and closets for woolens. And let us not forget the tanners. Oak and hemlock bark was used for tanning leather, which furnished everything from hinges and harness to shop aprons and shoes. Leather was the sinew that stitched the wooden age together, and it derived from the forest almost as much as from the cows.

Many of the earliest towns sought to protect these critical wood resources by regulating their use through the commons system. Only local citizens, or proprietors, held grazing, wood, and timber rights to the town commons, rights that each individual exercised under the jealous eye of his fellow townsmen. Because many of the trees that grew on the commons, particularly oaks, were needed for the local economy but also had market value, there were frequent controversies about theft and misuse. Many towns passed regulations attempting to control how trees were cut, and men were warned or fined for breaking these rules. There are records of repeated attempts by the town of Ipswich to prevent profiteers from raiding oak trees on the commons for the export market; and on the other hand to constrain commoners from cutting down and wasting valuable timber trees merely for the firewood in their tops. Watertown (the block from which Weston was chipped) recorded similar friction over timber felling on its commons.[9] Such tensions had long been accommodated within commons systems, but that was not the tenor of the times. By the end of the seventeenth century, the great bulk of the land in virtually all New England towns had been distributed into the hands of individual owners to do with more or less as they pleased. We need to examine how well this policy of privatization succeeded over the long haul, especially when it came to protecting the common, overarching benefits of the forest.

For the next century or so, the forest did not fare so badly. By the end of the colonial period in 1775, woodlands in the older towns were much diminished in extent but still comparatively intact. Between one-quarter and one-half of the landscape was still forested, even though most towns were crammed as full of farms as the colonial agricultural system could support. The persistence of forest stemmed from two practical limitations on clearing more land for farming. One was that the local wood supply could not be reduced beyond a certain point, and the other was that the local cow herd could not be increased. The forest was protected by practical constraints on how much land could be brought into cultivation by the colonial agricultural system, which, as I explained earlier, rested on meadow hay. At the same time, woodland itself remained a vital part of an agrarian economy that relied heavily on local resources. Shipping of ordinary wood and timber overland was in most cases uneconomical, so it was not easy to replace these resources through long-distance trade. This gave the forest in each town a measure of value and protection. As forest became scarcer, stripping land of its trees to create farmland often reduced the value of the property, instead of improving it. Most farmers and villagers alike owned one or more woodlots, and local artisans also relied on nearby forests to supply their raw materials. These woodlots, often detached from the home farm, tended to be found clustered in swampy areas and in rocky, hilly parts of towns that were least suitable for cultivation. They were thickest in the back corners of towns, where they adjoined similar country in neighboring communities, creating extensive forested tracts of hundreds or even thousands of acres. Walden Woods, lying between Concord and Lincoln, is one famous example. The limits of a balanced agrarian economy worked to keep a good portion of the countryside covered with trees for almost two centuries after Massachusetts was settled.

All this changed with the commercialization of agriculture during the first half of the nineteenth century. The number of farms did not change much, but the landscape changed dramatically. The spread of improved pastures and cultivated hayfields over the uplands to support more cows and meet the rising urban demand for dairy products led to the rapid decline of forest in places like Weston, just as the wool boom drove a spectacular wave of clearing in the younger hilltowns of up-country New England. This conversion of the remaining woods to farmland would have been disastrous if people in the older towns had still needed their forests for timber and fuel, but of course they no longer did. In Weston, as a consequence, pastures expanded and forests shrank rapidly. Firewood grew scarce and was in high demand as a quality fuel, but the high price simply served to lubricate the labor of clearing rather than to protect the local supply. It was not merely the increased demand for wood products that depleted forests in Weston—it was the availability of imported substitutes to replace local wood supplies that left the forests wide open to agricultural clearing.

What made this elimination of woodlots possible was the advent of coal as a cheap source of heat in the 1830s and 1840s as well as the ability to bring down wood from up-country by rail. About the same time, buildings constructed of massive frames hewn from local timber in the traditional way became scarce, while structures built by nailing together the new balloon frames made of smaller members of sawn lumber shipped in from Maine proliferated. Many small local tanneries went bankrupt during this period and were replaced by larger operations near the big cattle market in Brighton on the outskirts of Boston, which bought their tanbark from commercial producers on the frontier. The Hobbes tannery in Weston made the transition to the commercial era and survived for a few decades, but its tanbark came from Maine. Not all use of local forests for firewood, timber, tanning, and the like ceased, but protection of a local supply of forest resources became economically unnecessary in places like Weston.

The economic transformation was devastating to the forest. By the middle of the nineteenth century forest cover in Weston and many other towns in eastern Massachusetts had been reduced to only about 10 percent of the landscape. The long-term ecological impact of this disappearance of forest has hardly been studied, as far as I am aware. When we worry (as we should) about forest fragmentation and its effect on interior nesting birds, vernal pool dwellers, and other forest species in our own time, we have to wonder what was happening to these creatures a century and a half ago, when forest throughout much of the East was much more drastically broken up. We are lucky that so many species survived to repopulate the forest when it returned. Of course, some didn't, most notably the immense flocks of passenger pigeons that once swept across the landscape, carrying great loads of nuts in their crops and coincidentally moving tree species to new locations. And we know very little about changes in the populations of many herbaceous species of the forest floor. The trees have thankfully returned, but how ecologically robust these forests are beneath the canopy is another question. It is certainly one that we need to be asking ourselves as we go about caring for the forest.

By the mid–nineteenth century both the forest and much of the native wildlife that had accompanied it were in steep decline, both from loss of habitat and from overhunting. This was the time of the jeremiads of Henry Thoreau lamenting the loss of wild things. His house at Walden Pond was located in one of the few surviving extensive woods in the area. Thoreau walked through a degraded landscape, a rural world not yet overbuilt but seriously overfarmed. One unforeseen result of this agricultural expansion was a dramatic increase in summer flooding along the Concord River, caused by the deforested watershed's dumping much more water straight into the streams following heavy rains. The landscape was seriously disturbed and degraded by the loss of forest, top to bottom. Even in a natural world that is full of change, we cannot take any course we please without paying the consequences. By 1850, the maximizing of the land's output of a few profitable agricultural commodities had decimated the forest and created a patently unsustainable ecological situation in New England.[10]

Why did this happen? In New England, at least, it was not a simple matter of a continuous onslaught from the moment the Europeans arrived, but more of the abrupt release of economic drives that had previously been held partly in check. During the colonial era, greed had been constrained by an economy grounded firmly in local ecological limits. When these limits were broken, the forest became economically expendable. The utilitarian intimacy with the forest that had grown up during the age of local forest reliance did not generate a "forest ethic" sufficiently strong to protect the trees against the rampant demand for more farmland that developed in the early nineteenth century. Converting forest to cropland satisfied a powerful cultural ideal of land improvement, of course, but this drive had not been sufficient to reduce the forest beyond a certain point in an earlier economy centered on local community subsistence. The rise of a commercial approach to the land and its products drove forest to the brink.

Even in the colonial period, when local forest products were materially necessary, the other common benefits enjoyed by leaving enough forest in the landscape had never been sufficiently recognized or accorded adequate value. These priceless gifts of the forest include climatic and hydrological regulation, biodiversity, and beauty. To whatever extent our forebears may have recognized and cherished these benefits, no effective mechanism existed for protecting them within a system of private property. This possibility was precisely what had been given away when the commons system was abandoned within the first few generations of settlement. Of course, there is no guarantee the forest would have remained protected as commons, either, but as private property it lay completely defenseless. For a time, the common benefits of the forest were to a large extent coincidentally maintained because woodlots and their products retained sufficient economic value to their many private owners. But there was nothing to safeguard the forest's value to the community as a whole when a growing commercial market for farm products undercut the forest's profitability to those who held title, piece by piece, and so the remaining trees fell like ripe hay to make room for more cows.

Alarm bells began to ring during the early decades of the nineteenth century as New England's forest rapidly disappeared. Agricultural observers warned not only of the improvidence of losing regional self-sufficiency in wood products, but also of advancing signs of ecological (of course that word did not yet exist) degradation such as a hotter, drier climate and erratic stream flow—severe floods alternating with streams that dried up to nothing. Such observations were often muddled or exaggerated, especially regarding climate. On the whole, however, these pioneer environmentalists were right to be alarmed. The culmination of the early movement to protect forest came with George Perkins Marsh's *Man and Nature,* published in 1864. By then Marsh had been writing about the importance of the forest in popular agricultural journals for several decades, as had many others. The first cries of "Woodsman, spare that tree!" were heard in the 1820s, becoming a chorus by midcentury. These writers remonstrated with their contemporaries, chiding them that while clearing forest had been necessary in the early days of settling the country, it was now time for Yankee farmers to lay aside their axes and restore a better balance of woodlands to the landscape. Their aim was to instill a new respect for forest as part of an improved agrarian way of life and a more stable rural landscape.

One of the most eloquent spokesmen for the forest was George B. Emerson, who published his *Report on Trees and Shrubs of Massachusetts* in 1846. The book was widely read and well appreciated by the likes of George B.'s distant cousin Ralph W. Emerson and his young tenant Henry D. Thoreau in Concord. Although *Trees and Shrubs* certainly cannot match *Walden* in grace or depth, it ought to be celebrated nonetheless for the clarity of its early vision of the necessary double bond of affection and responsibility between people and their homeland. In a passionate foreword, George Emerson cataloged the many benefits of the forest and called for its restoration, urging that suitable native trees be replanted on every farm from Cape Cod to the Berkshires. He recognized the forest's economic value, he recognized its ecological value, he recognized its esthetic value; and above all he recognized that ultimately these values are inseparable.[11]

For George Emerson, restoring forest was a matter of cultural renewal and survival. Without it, he foresaw the continuing decline of rural New England towns that was already under way, as young people were drawn off to new urban employment and to fresh agricultural lands in the West. By midcentury, much of the immediately available forest and soil resources of upland New England had been rapidly depleted in the drive to market farm products. The land appeared to be run down and used up. The ravaged forest and ruined pastures proclaimed as much from every hillside. Indeed, there was a sense of urgency and impending doom in these warnings that had moral as well as ecological roots, a fear that New England was going to the dogs of worldly materialism. It is a lament that is as old as New England itself, and it is still with us today, I am proud to say.

This noble effort to convince New Englanders to cherish and nurture their forest largely failed, but the trees recovered anyway. The voices of such prophets as Emerson, Marsh, and Charles Sargent were important on a national level because they helped launch the conservation movement of the late nineteenth century. Subsequent efforts to protect forest within the region also descended from them, and their ideas were not forgotten. When it came to deliberately replanting New England with trees, however, these crusaders left virtually no trace. A few gentleman farmers planted trees on their estates—Nat Marden's grandfather planted pines on the land beside Fairhaven Bay in Concord around the turn of the century, for example. Earlier, Henry Thoreau planted a few white pines in his old beanfield on Ralph Emerson's land out in Walden Woods. As it turned out, a railroad fire burned those trees out a few years after his death. Others grew up in their place. Another, much more powerful restoration movement was under way that had more to do with what New England farmers weren't doing with their land than what they were. Having worn out their pastures and discovered western grain, they were no longer farming as much land of their own.

Fortunately, the forest still had the regenerative capacity to restore itself, or a new version of itself. Even at lowest ebb, woods covered enough of the landscape for seeds to constantly shower the encroaching farmland. A new crop of seedlings sprang up between the stone walls every year, ready to take over should the farmer's vigilance flag. It is probably just as well that nature and not people chose which trees to plant—I have little confidence that even the most enlightened lovers of trees would have chosen so well. As New England pastures and fields were rapidly abandoned in the second half of the nineteenth century, they were overgrown primarily by red cedar and white pine in southern regions and by red spruce and white birch in the highlands. This was nothing new: pastures had long been reverting to forest in older New England towns like Weston, even as new land was being cleared. Because the rate of clearing had accelerated in the early nineteenth century, there was a corresponding spate of pasture abandonment toward the end of the century, as the upland gave out. Forest reclothed about half the landscape by the end of the nineteenth century, farming contracting geographically even as it was still expanding economically.

This sea of dark green pasture pine gave an entirely new face to Weston and other southern New England towns and also gave rise to a new forest industry. White pine had not been as common as pitch pine in the native forest of eastern Massachusetts, but in the wave of reforestation following 1850 it expanded vigorously at pitch pine's expense, taking over virtually all the open ground except Cape Cod and a bit of the South Shore around Plymouth. This was a simple result of the suppression of fire. Burning had favored pitch pine on sandy soils for thousands of years during the period of native land use and during the centuries of European clearing. After 1850, fires were suppressed mainly by the simple expedient of not setting them anymore. The resulting upsurge of white pine led in turn to a pine lumber boom that lasted from the late nineteenth century until about World War II. Some of the pines provided construction lumber, but for the most part pasture pine was cut for box boards. Pine boxes were used to crate apples and to ship all kinds of goods, from machinery to mackerel. Eventually, cardboard and plastic packaging replaced pine, and after flourishing for half a century the box board industry faded away.

When the economic decline of New England farming began in earnest between the world wars, more land fell out of agricultural production. A second wave of reforestation began, bringing Massachusetts to about 65 percent forest cover by midcentury. More farmland has returned to trees since then, but this increase has been roughly canceled by forest cleared for development. Many of the white pine stands that sprang up on old pastures in the early decades of this century have never been cut, and like the stand on the college land are gaining impressive stature. Some older pasture pine stands that were cut long ago have grown up in pine-hardwood mixtures with an abundance of red oak, and old oak woodlots are reaching maturity. Red maple now dominates the once-drained hay meadows in the low places. This is a diverse and beautiful forest in which we feel quite at home, but it is not the forest that greeted our culture. Furthermore, it is never going to become that forest again, no matter how long we wait: it has no notion of a climax state. No Westonite of any race from any previous age would recognize today's forested landscape, dominated by white pine, red maple, and red oak, as home. It is the particular forest that has emerged from our own history, and it is up to us to figure out how best to care for it.

What sort of management of this forest, if any, is warranted? During the early part of the twentieth century the movement to restore, protect, and manage New England's forest was led by foresters, but in more recent decades many environmentalists have grown wary of that profession. This has to do with the ascendancy of the wilderness movement, and the association of tree cutting with everything from mild managerial arrogance to gross ecological plunder. Unfortunately, foresters were slow to incorporate a holistic, ecological viewpoint into their mission, although many of them have been making amends in recent years. Scientific forestry in southern New England grew up in the shade of the pine industry of the early twentieth century. The seedling profession adapted itself to pine as the region's most valuable timber crop and barely tolerated most other trees. Tree planting efforts usually featured white and red pine—in Weston, we still have a handful of these regimented stands of red pines, half a century old and never thinned. The need for cordwood as fuel was gone, no great demand for hardwood lumber was foreseen, and so hardwoods were denigrated, especially on sandy soils. The depravations of gypsy moths and chestnut blight in the first decades of the century cast doubt on the future of oaks and chestnuts as valuable timber trees, and in any case, hardwoods grew slowly. Foresters felt pressure to show public and private landowners alike that protecting and managing forests made good economic sense, and pine seemed to be the best answer to this challenge.[12]

This narrow focus gave forestry a bad name among many environmentalists, and for good reason. A story illustrates where forestry went wrong. When Weston's first Town Forest lands were acquired in the late 1950s, many foresters were still advising landowners to eradicate oaks in pine stands. This was actually tried in part of Jericho Forest in the early sixties. Hundreds of oaks were girdled and treated with herbicide. As a demonstration of policy for a community forest, this was a disaster—Rachel Carson's *Silent Spring* was just published. To make matters worse, the job was botched, and many of the wounded trees lived on in a crippled state. We sometimes lead walks through that part of Jericho today, showing people the bent old oaks among the pines, the encircling scars still painfully visible. Incidents such as this reinforced the perception that foresters did not see the forest as a whole, but only as so many board feet to be fattened for the teeth of the sawmill. Forest management was repudiated by the newly formed Weston Conservation Commission, led by Dr. Elliston. From the 1960s on, many forest owners began to look at their woods less as timber producers and more as wild sanctuaries and ecological buffer zones around overdeveloped housing tracts and shopping malls, and forestry was often seen as not only unnecessary but inimical to these values. To this day, in our region most town conservation commissions, land trusts, and environmental organizations that own land have no taste at all for managing their forests.

The monoculture approach to forestry turned out to be economically shortsighted, as well. Tree crops take a long time to mature, and the world changes. While the foresters were laboring to grow pine, the market for pine boxboards vanished after 1940, and the market for pine lumber softened. Meanwhile, things began to look up for hardwood. In the 1970s the firewood boom unexpectedly made low-quality hardwoods worth the trouble of cutting, at least. Then in the 1980s the price of high-quality oak timber went sky high because of the popularity of open-grained furniture and high demand in Europe and Japan, while the price of pine languished so low it was seldom worth cutting. So much for suppressing oak in favor of pine—perhaps the pines should have been strangled instead. In the 1990s, the price of pine finally recovered, partly because of political restrictions on excessive clearcutting in the Pacific Northwest, which I heartily support. Who could have predicted any of these shifts, and who can tell what will come next? The prevailing attitude among good foresters today—and our philosophy in Weston—is to bide our time and to manage for diverse, healthy stands of different species suited to particular sites, for both ecological and economic reasons.

Many thoughtful environmentalists have drawn a very different conclusion and wonder why management is required for any reason at all. New England was naturally forested and agriculture was never appropriate for very much of this land of marginal soils and harsh climate, they argue. This was amply demonstrated by its demise. Forestry has proven to be equally misguided because it is a similarly inept imposition of narrow human demands on the natural order. By its own regenerative vigor, the forest has rebounded in spite of our worst efforts. Why not just let go of the idea of human management altogether, let nature take care of itself, and cherish its wildness? In time, we will have old growth again, if we can only keep our hands to ourselves. This preservationist philosophy has guided much of the conservation movement in New England since the 1960s.

This is an inadequate view of the value of forest, in my opinion. It is a case of not seeing the trees for the forest. It declares that the most uplifting places for humans on this earth are those in which we can avoid having any functional connection with nature at all. But we must have a functional connection with nature to live. Most environmentalists, especially in places like Weston, are affluent people whose consumption of forest products is large. How can this be reconciled with the idea that we should refrain from managing our own forests productively? Why should we enjoy this luxury, unless we baldly state the truth that we would rather such unseemly extraction take place someplace else, out of our sight? It is all very well to take the position that we should drastically reduce our wasteful consumption of all the earth's resources and thus take the pressure off the forests. I agree, although I don't see many of my fellow suburban environmentalists out in front making such a change in their own lifestyles. But if we carefully examine the implications of such a desirable change, would it mean that we would become *less* dependent on the local forest? Quite the contrary.

The only way to protect the forests we love, in the long run, is to learn how to cut their trees responsibly. The forests of New England are again under assault. The worst threats come not from improper logging (although there is plenty of that), but from fragmentation by sprawling development, a steady bath of air pollution, and the onset of rapid climate warming. Wilderness preservation cannot protect trees from continental and global-scale degradation of the air that surrounds them and passes through their bodies, as it does through our own. To seriously address these problems, we must make enormous reductions in our consumption of oil and coal and in the way we routinely transform and transport materials around the world. It will also mean a reduction in the way we ourselves move around. One likely consequence will be greater reliance on sustainable use of local resources.

I have already argued that when the subsidy of cheap fossil fuel is withdrawn from industrial agriculture, we will need to restore a more robust regional food system in New England—but *only* while leaving a continuous matrix of intact forest covering a good half of the landscape. This protected forest should include most wetlands and riparian areas and also the steepest and least farmable of the uplands; but it should connect through some better soils in the middle part of the landscape as well. Such a permanent forest cover would protect biodiversity, but it would also provide wood and timber. The arguments that favor reducing our dependence on fossil fuel and on abusive industrial exploitation of land in distant regions to supply our food apply just as strongly to the way we obtain forest products. Making careful use of locally available, renewable forest resources should be seen as an environmental necessity. If we love the forest, we cannot afford to simply leave it alone. It is our responsibility, and it should be our joy, to restore some of the daily intimacy with forests and wood that existed here two centuries ago. This means cutting trees.

I can illustrate the paradoxical idea of cutting trees to protect the environment by a small example. Cutting and burning more trees in New England could help slow the buildup of carbon dioxide in the atmosphere and thus reduce global warming. This seems contradictory on the face of it—after all, we are often urged to plant trees to soak up carbon and cool the planet, and for sound reasons. We certainly need more trees in our cities, and planting them there is all to the good. We certainly need more trees on a global scale, and there are many regions in the world where shattered forests ought to be restored. When large areas of forest are clear-cut and burned and converted wholesale to pasture, the result is a net global increase in carbon dioxide. The case is quite different when a tree is cut for firewood within an established, well-managed, healthy forest in New England. The wood produces carbon dioxide when it is burned, to be sure, but meanwhile back in the forest other trees grow to fill the hole left by the tree that was cut, turning carbon dioxide back into wood. The effect is the same as if the tree had died and rotted in place, and in the end the direct net change in the amount of carbon dioxide in the air is roughly zero. In other words, we could go on cutting and burning wood in this way forever without measurably affecting the climate, because we'd be operating within the current carbon budget and keeping the balance between atmosphere and vegetation about the same. Meanwhile, by heating a house with wood instead of oil we replace an equivalent amount of fossil energy, thus reducing the rate at which long dead and buried carbon returns to the atmosphere. By burning one or two gallons of gasoline in a chainsaw and a truck, we can deliver a cord of solar energy equal to 150 gallons of fuel oil to the door, keeping the other 148 gallons safely underground for another season. Cutting trees to prevent global warming is an excellent syllogism for a sustainable, solar society.

The same is true of everything else that we could be harvesting from forests in New England but now import. Consider lumber. A house that is not built of local white pine is built instead of lumber shipped all the way from the clear-cut old growth of the Pacific Northwest or the monoculture plantations of the South. Or perhaps it is built of concrete or steel, in which case some environmentalists rejoice because trees have been saved. But the tremendous increase in building with aluminum, steel, and concrete in the twentieth century and the relative decline of wood construction is mostly a function of cheap fossil fuel. These mined, smelted, and manufactured earth materials are much more energy intensive than lumber, particularly locally harvested lumber. Roofs built of steel rafters instead of pine two-by-eights may save trees right now, but in the long run they do not save forests. If we do not shoulder the responsibility of producing what we can from our own forests, sustainably, we face the choice of buying the same products made from wood grown elsewhere, using nonwood substitutes, or simply doing without the thing in the first place. In some cases doing without is the best choice, in some cases the nonwood or imported wood product has a special quality that justifies its use, but in most cases the local wood product is the most satisfactory. I am not suggesting we eliminate metal and plastic, but we can surely get along with far less of them and with proportionally more wood. We New Englanders live in a forested region, and our homes and furnishings ought to reflect it, at least in rural areas. Our forest provides some of the most diverse and beautiful woods in the world, as our ancestors knew. As New England environmentalists, we should, in general, prefer a table of oak to a table of teak; a house of white pine to a house of Douglas fir; a woodstove to an oil furnace. We should feel good, not guilty, that trees have been cut to produce these things. But we must learn to manage our forests sustainably, in an ecologically justifiable use of that term.

Principles of Sustainable Forestry

As we at Land's Sake continued our modest firewood cutting program through the 1980s, we faced a challenge: how to introduce the radical idea of timber harvesting into the cherished conservation forests of a suburban town. Firewood thinning was getting people used to the rattle of chainsaws in the woods, but what was the next step? After the near fiasco of logging the college pines, we knew that the woods in Weston had to be both marked and cut in an extrasensitive way, or they would not be cut at all. We also realized that, first, we needed to have a comprehensive plan for the management of town forests that would put logging operations in a long-term context and, second, that we would have to conduct an ongoing public education effort as we went around felling trees in the woods behind people's houses. By the end of the decade, the Conservation Commission and Land's Sake had put those pieces into place.

We acquired a Farmi logging winch, made in Finland, to run on our new John Deere 2155 tractor. The winch was powered by the tractor's rear power-take-off and was equipped with a couple hundred feet of cable to pull felled trees from the forest to landings along the logging roads. Once pulled in and hooked to the winch, logs could be raised and skidded farther along by the tractor if need be. The process wasn't quite as efficient as having an actual skidder, but that little unit has pulled some impressive logs out of the woods. The tractor, winch, and a succession of beat-up one-ton dump trucks made up our basic logging gear.

For forestry, we did even better. John Potter returned from Yale with a progressive forestry education, and his classmate Jeff Bopp joined him to give us a full-time forest team. John had started splitting wood with us during our first job in Jericho Forest when he was about sixteen. For his master's thesis John conducted an inventory and prepared a set of long-term management goals for Weston's forests. We presented this plan to the Conservation Commission, they accepted it in principle, and we got to work with a clearer vision of what we were doing in the woods.

John Potter's forest plan included what he termed ecological protection, wood production, education, and recreation. These amounted to a forester's version of Land Sake's four familiar interlocking and reinforcing principles: ecology, economy, education, and esthetics. Ecology comes first, as always. The key to the plan is that it treats Weston's approximately fifteen hundred acres of accessible town-owned forest land as a unit, even though this forest is in fact scattered across town. Most of the forest we work with is concentrated in a few large blocks, but we treat everything together as a conceptual whole called Weston's town forest.

ECOLOGY

Land's Sake's ecological principle is to manage the town forest for diversity of both species composition and age classes. John's survey confirmed that Weston's forest is divided into numerous small stands of red maple (31 percent), white pine and pine-hardwood mix (33 percent), oak-dominated hardwoods (35 percent), and a handful of hemlock groves. Most of these stands cluster between fifty and eighty years old—middle-aged, just coming into their own. Weston has few young stands anymore and little old growth as of yet, which is typical of the region. Almost all of the present forest was regenerated by a series of events that occurred between the two world wars. During this period, pastures and hayfields were abandoned and woodlots were cut for the last time as both working farms and large estates faded from the scene, and the land was left alone to grow trees. The chestnut blight destroyed stands dominated by this wonderful tree in the early 1920s, allowing other species to come on—in one part of the Highland Forest, a lovely stand of white birch was the result. Finally, the Hurricane of '38 flattened many maturing stands of pasture pine. In some of these pine grew back, while in others young oak seedlings were released, resulting in an oak-pine mix. One way or another, almost all the standing forest in Weston dates from this period.

If nothing at all were done, this forest would probably mature into magnificent old growth of tall, stately trees over the next century or so. Probably. At some point, another hurricane would come along and destroy a large part of this generation of trees in just a few hours: a perfectly natural event that, as far as we can tell, happens upon a significant swath of southern New England every century or two. Less catastrophic natural disturbances including fires and disease outbreaks affect smaller areas on a more frequent basis. Tornadoes always seem to head straight for revered stands of old growth. As a result, over the entire region in the past there have probably always been some young, regenerating patches, some middle-aged patches, and some mature forest, making up the shifting mosaic. In the decades after a big blow, the prevalence of younger stands greatly increases in the affected area, while mature stands reach their full development elsewhere. Wildlife adapted to each of these conditions can find places to live throughout the wider region, with the populations of various species increasing in one area and decreasing in another, responding according to their requirements. For example, robins, song sparrows, and many of the familiar birds of farmland and suburb today once flocked to the young, open areas that followed disturbances. Goshawks and many neotropical migrant songbirds, on the other hand, flourished in patches of forest that had become more mature. The deer went one way, the squirrels the other. Left to its own devices, nature would in time restore a broadly similar landscape of alternating mature and young patches across the region, but these alternations might take place in an abrupt manner that suburbanites would find disturbing.

Mature trees are more susceptible than younger trees to lethal disturbances, especially blowdown, for obvious reasons: they stand higher and more exposed, and the leverage the wind exerts on their lower trunks and roots is enormous. Allowing mature stands to dominate over a wide area invites the day when nature will with benign indifference instantaneously replace a large part of the forest with stands of seedlings a few feet high, threading their way up through a ten- or twelve-foot-deep litter of fallen giants. This is not an ecological disaster. The huckleberry tribe and its retainers rejoice. Such species as birch, aspen, and fire cherry, which relish such opportunities, usually have a toehold somewhere nearby and the capacity to travel great distances and spring up over wide territories in short order. Most of the species that will eventually make up the mature forest grow alongside these weedy pioneers, only a bit more slowly, and gradually reassert themselves.

The question is, Do any but the deepest ecologists really want to occupy houses in the midst of such monumental natural topplings of tall trees and the roaring slash fires that often follow in their wake a few years later? Probably not. We do not inhabit a sparsely populated forest region anymore, but a densely populated one in which forest and people who invest in fixed property coexist in close proximity. For this reason alone, we face the challenge of managing the forest in ways that mimic natural patterns and accommodate natural processes but that also allow for our own presence.

One way to do this is to attempt to reproduce within the smaller boundaries of every town something of the diversity of patches and species that would naturally be spread over a wider region. That is, we can intentionally scatter smaller-scale disturbances throughout the landscape on a regular basis. This constantly recreates young, regenerating stands, keeps a large part of the forest in the middle-age classes, and (provided the cutting rotation is long enough) ensures that a good part of the forest will always be found in ecologically mature stands, as well. Thus habitat always exists locally for the widest possible variety of forest species, especially if we see to it that unusual species and ecosystems are protected and encouraged and that a few larger blocks of mature forest always remain somewhere within the region. Such measures are also disaster insurance against the day a hurricane blows through because a larger percentage of the forest will escape having its calendar reset to the year zero and surviving middle-age ecosystems will become mature within just a few years. Then when a stately old grove topples, as recently happened in Connecticut, it will be no great calamity.[13] Such a forest is certainly not purely "natural," but then again purely natural forest has not existed in the region anyway during this interglacial period. Repeated across the landscape in town after town, however, such forests should provide ample opportunity for all natural species either currently residing in the region or headed our way to flourish far into the future. Land's Sake and the Conservation Commission aim to create this opportunity in Weston.

The idea of making disturbances that resemble hurricanes no doubt strikes many environmentalists as suspiciously like the rationale the logging industry often uses to justify its massive clear-cuts. As a matter of fact, clear-cutting is an appropriate silvicultural technique for regenerating certain highly shade-intolerant species such as white birch, something else industry apologists are fond of pointing out. On the whole, however, Land's Sake's philosophy is strikingly different from industrial forestry as it is now practiced. We might share some of the same rhetoric, but we really mean it—and don't mind taking people on walks in the woods to prove it to them. The most telling difference, perhaps, shows in the length of the rotations. Industrial forestry turns the forest over as rapidly as possible, often breeding fast-growing species expressly for this purpose and even spraying herbicides to eliminate competition. The day she catches me spraying poison in the woods for some supposedly benign ecological purpose is the day I want the local arboreal goddess to loose the great goshawk on me—I have had enough of that. Our approach is to manage for old growth, with projected rotations approaching the length of the hurricane cycle, about 150 years. We encourage species diversity and insist on forestry and logging that pay careful attention to the details of each stand, instead of narrowing the species of interest and simplifying the landscape to suit uniform management practices. The idea is to work with natural forest patterns and rhythms, not to create oversimplified plantations and blow them down methodically every few decades with feller-bunchers, calling it the moral equivalent of a hurricane. In other words, Land's Sake puts ecological goals first, instead of concocting ecological excuses to justify practices that are really determined more by rates of return on investment and business cycles than by natural cycles.

The way Land's Sake makes its regenerating cuts exemplifies this difference. The object in sustainable forestry is to create conditions that resemble those under which most of the species in a particular stand naturally reproduce and to which they respond best on their own. It is not necessary to plant trees to replace those that are cut—this is more pabulum. In fact, planting seedlings is usually a sign of bad forestry. The environmentally warm, fuzzy image of tree planting gives the big timber companies an easy public relations opening to brag about the billions of trees they plant, which generally means that they are busy turning thousands of square miles of former forest into monoculture plantations. Natural forests replant themselves constantly and spring back on their own after a stand-opening disturbance of some kind comes along and releases this advance regeneration. Many of the trees that now dominate the forests of Weston, for example, red oak and white pine, reproduce very well in partial shade. This creates an opportunity to use a method of regeneration cutting called shelterwood harvesting.

Shortly after returning to Weston in 1989 to work full-time with Land's Sake, John Potter introduced us to shelterwood cutting. We took on a job harvesting timber in a six-acre private woodlot. The stand was dominated by red oak nearly two feet thick, along with white ash, red maple, and an upgrowth of younger, shade-tolerant sugar maple that had seeded in from the road over the past century. As we got into the cutting we discovered that many of the trees were 140 years old, trees as venerable as any I have seen in Weston. It was a fine old piece of woods that demanded first-class treatment. Nine years before, when I picked up a chainsaw and started Land's Sake's forest program, cutting that woods would have been out of our league. Not anymore.

John engineered a shelterwood "removal cut," designed to be the first of several that will eventually harvest and regenerate the stand. In shelterwood harvesting, the timber crop is taken not in a single clear-cut, but in a series of two or three heavy thinnings spaced over several decades. The first "prep cut" sets the stage for regeneration by removing a third to a half of the standing timber, encouraging the remaining mother trees to produce seeds and shelter their offspring. The cut is heavy enough so that the canopy cannot close and shade out the small seedlings that always struggle underneath, as it does in the routine firewood thinnings that Land's Sake carries out in younger stands. Instead, enough light continues to reach the forest floor year after year so that seedlings begin to shoot up rapidly in the partial shade of their parents. In the meantime, the remaining timber trees are no longer surrounded by the close neighbors that forced them to grow tall and straight, but can spread their crowns and swell their trunks with wood. When these great trees are finally harvested a few decades later, the younger trees beneath them will be far past the seedling stage. Instead of a raw clear-cut, a flourishing young stand of pole-sized trees will greet the forest lover walking along the trail. Thus, along with its ecological merits, shelterwood harvesting yields economic and political benefits because the ground is never bare of trees. The old stand and the new stand overlap by a few decades.

That fall and winter we stacked up sixty cords of firewood on that six acres and skidded out twenty thousand board feet of oak and ash timber with the little John Deere—some of it veneer quality. But even better than what we took out of the woodlot was what we left behind: a beautiful stand of tall, soaring oaks, ashes, and maples, widely spaced and ready to fill out: a real showcase. I only hope that in twenty years this woodlot will still be protected and that Land's Sake foresters will still be around with an idea about how to make the next removal cut. And I hope that by the time the final cutting is made there will be many more mature stands of equal stature that have come along in Weston to replace it. The key to good forestry is continuity and long-term commitment.

You can't really practice good forestry in a single stand—you need an entire forest. As beautiful as the shelterwood harvest was, such forestry makes sense only in the context of a pattern of rotational cutting over a much wider area. This is exactly what we have to work with in the town forest. John's inventory identified 1,440 acres of accessible forest covering most of Weston's conservation land. To provide a natural refuge and a standard for our manipulations, we proposed setting 10 percent of this town commons aside as a permanent wild reserve. As we carry out a biological inventory we expect to find unusually sensitive species and ecosystems that require special protection, and in some cases special management. This will leave us well over 1,000 acres to work with, a forest already composed mostly of discrete 5- to 20-acre stands separated by stone walls, ready-made for rotational management. If 10 or 12 acres are harvested every year by the shelterwood method, it will take more than a century just to work around the town for the first time. By then, a more even distribution of age classes will have been slowly generated in our wake, and our grandchildren will be ready to start around again in their golden years. In the meantime, we and they will have routinely thinned young and middle-aged stands every fifteen years or so to promote the growth of the best trees, taking out firewood, pine pulpwood, and small sawlogs. This is the basic forestry program toward which Land's Sake is working in Weston.

Shelterwood harvesting is not appropriate for all species. Regenerating some, such as white birch, would require small patch clear-cutting and possibly prescribed fire. If not given a few opportunities like this, white birch could be virtually lost from Weston—except as planted by landscapers in front yards. Shade-tolerant northern hardwoods like sugar maple, on the other hand, call for a more selective timber harvesting approach, creating uneven-aged stands. In other areas we might want to deliberately reintroduce species that have been lost, such as black ash, white cedar, and, perhaps someday, blight-resistant varieties of American chestnut. As the threat of climate warming becomes real, we may even want to introduce a few species whose range lies just south of our region, such as black walnut, tulip poplar, and chestnut oak—again, not as a wholesale change over the entire forest but as a way of hedging our bets against the continued propensity of *Homo sapiens* to change prevailing environmental conditions faster than many other species can adjust. The forest must be kept diverse in many different ways. We are not so naïve as to believe that whatever prescription we lay down today will be slavishly followed by our successors for the next century. Our program is flexible and conservative and leaves many options open.

Of course, there is more to ecologically sensitive forestry than trees and the way they are harvested. Ecological management begins with healthy trees, but it hardly ends there. In fact, it is important to retain a few rough, diseased trees and dead snags throughout the forest and to leave plenty of deadwood on the forest floor to provide nesting cavities for birds and bees and fodder for centipedes and fungus. Forests are complex ecosystems composed not only of trees but of the herbaceous plants, animals, soil organisms, and other creatures that live with them. Many of these species appear to be thriving in the recovered New England forest, but many others have vanished in the past or are vanishing in the present. For example, many ecologists believe that neotropical migrant birds are declining as their habitats fragment and they are exposed to predators and pretenders such as cowbirds and cats. There are also rare plants in the Weston forest we need to protect.

We file cutting plans with the State Forester's office, which under Massachusetts law automatically triggers protection for any rare or endangered species that has been discovered on or near the site by the state Heritage program. Land's Sake hopes to conduct a more complete biological inventory of Weston, producing an ecological baseline for a monitoring program as our forest program proceeds. We might find opportunities to add ecological restoration to our program, reintroducing and encouraging species that are missing or rare—not because we need to exactly recreate the ecosystems of the past, but because we want nature to be playing with a full deck in creating the ecosystems of the future. On the other side, we have the invasion of such exotic species as buckthorn and Norway maple to pay attention to. It is not enough to create a forest made up of diverse stands of tree species and hope for the best for everything else—our responsibilities extend beyond that, if we are going to cut timber.

Ultimately, the Town of Weston should have a resident ecologist as well as its forester. The two should work in harness. Half a century ago Paul Sears dreamed of the day there would be an ecologist in every county in America, and today many New England towns do indeed employ a conservation officer. Unfortunately, much of their time is consumed enforcing the Wetlands Protection Act, a worthy piece of legislation but hardly a comprehensive policy for protecting and managing ecological systems. I'm sure many of them would love to spend more time out in town forests conducting biological surveys. The proceeds from good forestry ought to help pay for good ecology. In this way biodiversity could be protected on common forests in town after town by local vigilance, rather than by the hodgepodge preservation of special habitats by outside agencies and organizations. In the end, a wide variety of species must thrive not stranded in isolated preserves, but throughout the landscape and in conjunction with productive use of the forest.

ECONOMY

Managing more than a thousand acres of diverse forest in this way may be based on ecological values, but in the end it is also a sound economic proposition. Local forests can offer small logging operations a steady, year-round diet of work and so supply a nice variety of wood products. All that is required to make it pay is a comparable variety of local outlets for these products, from first thinning to final harvest. The first challenge is to market the culls. Marketing hardwood culls is no mystery for Land's Sake: firewood is the bread and butter of our operation. Cutting firewood is not highly profitable to either Land's Sake or the town, but it does cover the costs of making improvement thinnings, which increase the value of the remaining timber. The sustainable production of Weston's town forest is at least five hundred hardwood cords a year—roughly the equivalent of seventy-five thousand gallons of #2 heating oil. That's enough to heat a couple hundred well-insulated houses—hardly all the fuel Weston consumes, but a potentially significant contribution to the local residential energy budget. In only slightly less suburbanized towns with fewer houses and more trees, wood could provide most home heat indefinitely.

Finding a comparable market for pine culls has proven frustrating because Weston is a long way from the pulp mills. The town is full of middle-aged white pine stands that ought to be thinned and of mixed stands from which it would be convenient to remove cull hardwoods and pine at the same time. Chips for a wood-fired boiler might be a good outlet for cull pine, if such a market could be found. What we need in Weston, I think, is a new boiler in a municipal building that can burn either wood chips or natural gas efficiently. Then, by having a large chipper at the brush dump the town could utilize pine logs that landscapers and arborists presently drive miles to dispose of, along with what Land's Sake could generate in the course of improving pine stands in the town forest. Such an arrangement might be economically marginal (until the price of oil goes up again), but it would solve several long-term problems at once. Until some reliable local market can be developed for small and crooked softwood, it will remain difficult to practice good forestry in pine.

The most valuable wood products to come from a well-managed forest are high-quality pine and hardwood timber. The beauty of sustainable forestry, with its emphasis on long rotations, is that trees reach their highest value, rather than getting cut off in their prime just because their rate of growth is slowing. In other words, ecologically sound forestry promotes quality instead of quantity in timber production. The best, clear wood gets laid on last, after the tree has reached its full height—in the case of white pine and red oak, in the decades after it has turned eighty or one hundred years old. Weston can profit from growing quality timber as long as someone nearby is prepared to pay top dollar and make proper use of such wood. The sustainable production of the town forest under this kind of forestry will approach two hundred thousand board feet of sawtimber per year. Given the wonderful diversity of trees in Weston, this is enough lumber to frame, floor, and furnish about a dozen fine houses every year—enough to keep a small town standing indefinitely.

In the long run, the economic keys to a good forest management program are seeing to it that a wide variety of the highest quality forest products are steadily harvested and simultaneously developing strong local markets for these products. At Land's Sake, we want to capture the added economic value and the powerful ecological resonance of retailing directly from the forest to local customers, the way we do at the farm. Better yet would be working the wood into fine things through small-scale wood industries of our own or supplying private woodworking firms. This approach has worked well with firewood. Land's Sake doesn't sell bulk logs to a processor. We work it up ourselves into top-notch firewood, employing young people in an educational and enjoyable way and selling on the high-priced retail market. Now we need to do the same thing with our pine thinnings and high-quality pine and hardwood timber.

For several years Land's Sake experimented in milling lumber with a small bandsawmill. John Potter's brother Michael cut beautiful pine and hardwood lumber with this mill. We sold some of it to local woodworkers, but mainly we used it for construction projects: building our farmstand, hay wagons, and trailers and putting a new hayloft in Dr. Earle's barn. Michael even borrowed a lathe and turned out white ash softball bats, which had sympathetic powers to drive the ball out of the field and back into the woods. All this was very satisfying but not profitable. The mill cut nice lumber, but it was too slow and inefficient to make money, and we eventually sold it. The experiment proved to us not that small-scale milling is impractical, but that we needed a more efficient professional model of sawmill. This would allow Land's Sake to build a local market for lumber produced by ecologically sustainable methods. We hope to return to milling lumber in the near future.

The short-term key to making sustainable forestry succeed is the same as in organic farming: find customers willing to pay a premium for superior local products. The superiority of a natural product lies both in the intrinsic quality that results from the care with which it is grown and crafted and in the added value that customers attach to it because it is produced locally by ecologically sound methods. Like the market for organic produce, the green forest products market will have to be developed by patient promotion. Buying such wood products will prepare people for the day when they become the norm because many of the cheaper substitutes will no longer be so cheap when we begin paying the full ecological cost. To hasten that day, environmentalists should take it upon themselves to seek out and support sustainable community forestry in the same way they do local organic farms.

EDUCATION

Weston's town forest is a commons. Land's Sake's approach to the forest must be endorsed by the community as a whole to be successful. It isn't just a matter of cajoling them into allowing us to run chainsaws in the woods for a few years—somebody is going to have to keep this up for decade after decade before it amounts to anything. Hundred-year plans are carried out by cultures, not individuals. Community forestry should seem as natural as running our own school system in Weston—in fact, it should be part of our children's schooling. Land's Sake doesn't just run formal school programs on behalf of the Conservation Commission: everything we do in the forest is designed to be educational, from start to finish. Learning about Weston's forests is a large part of community forestry.

Over the years we have moved our operation to different parts of the town forest, which means starting the saws and dropping trees behind a new neighborhood of forest lovers each time. Before we start cutting, we invite everyone interested to come for a Sunday afternoon walk, followed by coffee at the home of a Land's Sake member or conservation commissioner in the neighborhood. This is the way things are done in Weston. We spread the

Forest and Trail Association map on the living room floor and discuss the philosophy of the overall town forest management program and the objectives of the particular cut. Our neighbors air their concerns, and we learn things that we should know and so modify our plans. In one woodlot, for example, at the intersection of three trails stands a six-foot-high hemlock tree that someone decorates for Christmas every year. It is a neighborhood icon. Imagine the dismay had Land's Sake innocently wiped out that little hemlock to make way for our firewood truck! Community forestry means finding out who walks in the woods and taking a walk with them before cutting any trees. Not everyone will learn to love the sound of chainsaws, but you will have the weight of the neighborhood behind you.

Land's Sake has worked closely with Susan Majors, a biology teacher at Weston High School. Every fall we lead a unit of Susan's Environmental Science class on global deforestation and the plight of the tropical rainforest; we also discuss what can be done about this problem in Weston. Eating a thousand pints of Ben & Jerry's Rainforest Crunch is not enough—not that I have any objection to that, mind you. At the end of the week the class comes out in the woods to see how sustainable forestry and low-impact logging is done. Letting young people see that environmentalists can oppose ruthless clear-cutting and still enjoy dropping trees in the name of ecology sends a powerful message. We lead similar walks with local conservation organizations and set up self-guiding tours with signs along the trails where we are working. Taking care of the forest in Weston and harvesting wood products sustainably at home is not all that is needed to save the rainforest, but at least it starts to put our own house in order.

As always, the most important educational component of Land's Sake forest program is work. Most logging work is not suitable for teenagers, but woodsplitting is. It might be marginally more efficient for us to use a mechanical splitter, but handsplitting fits in well with our system of moving the wood the fewest times and the shortest distance between the tree and the delivery truck. Besides, it is more fun. A great virtue of splitting by hand is that it introduces young people to the forest in a way that is useful, satisfying, and memorable. Effective education is bound up with good economics, by involving young people in tasks in which the work and the products are real, not just exercises. This is not the only kind of education that is valuable, but it is an aspect of education sorely lacking in the usual suburban upbringing. The economy being practiced is embedded in turn in ecological values. These lessons are best taught with a six-pound maul.

After the wood is felled, skidded, and bucked to length, we split it and stack it between trees along the skid roads, where it can be easily thrown into the truck for delivery, come fall. Stacking the wood to season right where we cut it saves the additional step of moving it to a central location. The high school kids we hire to help split it come to work with us out in the woods; after all, what would they learn and how long would they stay slaving away on a mountain of logs behind some town garage? We would get a mechanical splitter fast if we had to handle the wood that way. I think the key to the entire forest enterprise is getting the next generation out in the woods. We're not just getting the wood split, we're raising future Land's Sake foresters.

We take kids out to split wood on weekend mornings throughout the winter. Not many people are effective woodsplitters for much more than four hours, believe me. We work to strike a balance between running an efficient operation and involving young people in ways that will turn them on. It's more cunning than strength, once you've got the basic stroke down, we tell them. Of course, a little strength never hurt. "You have to be able to hit it in the same place twice," I once instructed a novice woodsplitter. "Hit it in the same place twice?" she replied. "I can't even hit it in the same place *once!*"

Over the years Land's Sake has introduced dozens of Weston teenagers to the woods. A few become woodsmen (and woodswomen) in their own eyes, a breed apart, and they get to be smooth operators with a maul. As we work we notice the intricacies of the way different kinds of wood split, learn their colors and in warmer weather their smells. We talk about which trees we cut and why, the history and desirable fate of this particular piece of forest. Once they turn eighteen, the serious woodcutters are introduced to the chainsaw—first bucking, then felling. After a few years they can't wait to get home on their college breaks so they can get out to the town forest to cut and split wood.

Handsplitting is an extremely efficient way to process firewood, when all is said and done. A good woodsplitter can split wood in little more time

Jennifer Rossiter splitting oak, 1990. (Susan M. Campbell)

than it takes to set it on end. Seasoned woodsplitters know without thinking from which end and which angle to attack a log, how hard to swing, how to pace themselves. Good splitting looks effortless and feels effortless. Ordinarily, it takes as much time to stack the wood as to split it. A decent splitter can split and stack a cord in four hours, and on a good day some can do it in three. The working average for a season, what with walking in and out of the woods, digging around in the snow and finding runners, teaching new workers and all, might be closer to six hours. Above that we start losing our shirt. If I take a student into the woods for four hours on a Saturday morning and we get as much as two cords stacked up, I know we are making money.

By getting our wood split and stacked before the leaves come out in the spring, we have a reasonably combustible product to sell by the fall, which justifies the high price we charge. After the days begin to warm up consistently above freezing you can walk through the woodlot and smell all the oak drying, wafting down the slope from row upon row of woodpiles between the trees, as far as the eye can see. Delivery takes place in the fall, as the growing season winds down on the farm. By the middle of September, the farmer side of our nature dreads frost, but the woodsman side dreams of it. It's always a relief to get back in the woods after the intense heat and pressure of summer in the fields. Loading the truck and delivering wood is a pleasure: the logs are now dry and amazingly light and fly into the truck, our customers are pleased, and of course the money is rolling in hand over fist. Our wood is generally drier than most wood that is sold as seasoned these days because we stack it up so that air can get to it and don't just throw it in a big heap as many firewood dealers do. We also get an accurate measure, so that our cords actually stack up to 128 cubic feet by the customer's door. More than once I have dumped the first cord of a two-cord order and watched the customer write a check for the full amount.

"No rush," I say, "we'll be back in an hour with the other cord."

"Other cord? Where am I going to put it? The last guy I bought wood from gave me less than this and called it *two* cords!"

A retired physician was so delighted when we brought him twice as much well-aged wood as he expected that he gave us a well-aged bottle of Medoc from his cellar and became a Land's Sake member on the spot, contributing $100 a year ever after. Quite a payoff for good behavior. I always impressed this on the kids, although I did not share the wine with them, for legal reasons of course. Land's Sake is a community business that lives or dies by its reputation. If we don't set high standards, not only do we cheat our better natures, we get run out of town on a rail. If we do the right thing, people will support us. Within a community, people are more often than not done by as they do: to a large extent, you create your own ethical environment. Treat your neighbors as generously as you can, and for the most part your kindness will eventually be repaid, often in ways you could never have anticipated. The sharper kids also observed how cynically calculating we could become when we had to weasel our way through the tangle of town politics, so I think they got a well-rounded view of the peculiar mixture of warm sympathy and cold savvy that is required to get along in a community.

ESTHETICS

The woods must be beautiful. This is the final principle in Land's Sake's forest philosophy. The woods must be beautiful and the uses of wood must be beautiful, making a vital connection between people and the land. I am not talking about mere esthetics here, I am talking about how to create deep, enduring bonds. People should not only walk in the forest and enjoy its natural beauty, but be joined to it through their daily lives.

Recreation is often counted as the use of forests near urban areas that directly benefits the most people. I have no quarrel with that as long as it is seen as a beginning, not an end. In Weston, Land's Sake maintains more than sixty-five miles of trails for the Forest and Trail Association and Conservation Commission. These trails are enjoyed by walkers and joggers, horseback riders, cross-country skiers, and increasingly by mountain bikers. Weston's trails are not nearly as heavily used as they could be, however. They are inaccessible to those who do not live in the town and know the ways of Weston, while many who do live here are often away enjoying the outdoors somewhere else, instead. On a weekend you can probably find more solitude in the Weston town forest than in many better-known wilderness areas in New England. On a weekday you can be lost to the world. As much as I enjoy solitude, I think it would be healthier all around if more people were out walking in Weston's woods and other suburban forests. It is crazy to have millions of people driving great distances to love more spectacular scenic wonders to death, in the name of spiritual renewal. If suburban Americans were to take half the time and money they expend on wilderness vacations each year and spend it instead on preserving local forest lands and walking in their own woods, the world would be a markedly better place. Ecotourism begins at home.

Logging is compatible with recreational use of the forest, but it does take special care. At Land's Sake we don't run chainsaws on weekends. Trails are never left blocked with trees or wood (or even sawdust on snow, which trips up skiers—we learned that one the hard way); and of course we always use extreme caution before dropping a tree, anywhere. Cutting and skidding trees inevitably leaves a mark on the forest, but it can still look acceptable. Our skids are kept light and don't tear up the forest floor too badly, we are fanatical about not scarring residual trees during either the felling or the skidding, and we cut our slash so ridiculously low it virtually disappears beneath the horizon of natural debris. On the back of the John Deere's roll bar, up above the top of the Farmi winch, is a sticker that expresses our work ethic perfectly: "Cut Low Stumps." We leave the woods full of healthy trees, looking just a bit more open and airy than when we started. Within a few years the canopy closes, the skid roads fade away, and unless you really know what you are looking for, you can't even tell we've been there.

When the three fundamentals of ecology, economy, and education are all engaged, an esthetically pleasing result is virtually guaranteed. The artistic possibilities of working with the landscape do not end there, but that is their foundation. Without planning to, we created an unusual new kind of beauty in the forest, a form of woodland art. A century and a half ago, George Emerson urged Massachusetts farmers to plant mixtures of native trees with an eye to their form and color at every season, as a kind of full-scale landscape painting. This vision was remarkable because it suggested that a culture that understood and cared for its land would create a beautiful landscape, first unconsciously but finally deliberately, by continually reproducing those elements that provoked a familiar emotional response.[14]

Sadly, this noble idea has gotten lost in New England or has degenerated into tacky commercial nostalgia of the Ye Olde Shopping Mall at the Commons variety. As for forest art upon the landscape, except for some wonderful rows of roadside maples, nature itself did the grand repainting. Ecologically, this was probably for the best, but the result reflects a cultural loss because the living human patterns are covered over and gone. Woodlot management revives George Emerson's forest art, but in a different way. The landscape art we practice is less painting than sculpting: the artful subtraction of timber and firewood and the temporary installation of woodpiles.

The visual impact of our woodpiles surprised us. We made our piles in the woods for practical reasons, but when we were through, the place looked terrific. People loved walking through our woodlots, and they let us know it. Hundreds of small piles flanked the trails by the end of each winter, running at odd angles between the standing trees that buttressed them. Art scholars might explain the sense of serenity conveyed by these horizontal elements among the verticals of the thinned trees. They might point to the latent energy expressed by the juxtaposed curved and sharply split log ends embedded in the stacks and by their unpredictable angles. Each composition was "artless" in design, dictated by the discipline of building wood stacks that didn't fall over (too often), within the framework provided by the trees and by the rise and fall of the land underfoot. The stacked wood drew out all these things for the viewer. Above all, the piles illustrated that most of the trees were still there, alive and standing. The stacks neatly expressed our care for the forest. Our woodpiles were an unexpected artistic triumph and made us many friends long before we even met them.

A piece of firewood is a humble object, one lacking much artistic pretension—but it is beautiful. Wood splitting is a rudimentary form of sculpture, but it is an art, and one that is readily accessible. It is an enjoyable art to practice, combining strength and skill, and it is full of small technical solutions and sensual rewards. Woods reveal a surprising variety of bright colors and grain textures when split open, from tawny white oak and pale white ash to creamy red maple and wine red oak. Each has a characteristic smell and even a characteristic sound when it splits. I can often distinguish from far away the hollow pop of oak from the higher ping of maple or birch through the frozen woods. A good woodstack outside the door is a handsome thing, full of potential energy. Besides combustible hydrocarbons, all the pleasures of the woodlot are gathered in it.

Such artistry bursts forth again as heat and celebration when wood is burned. A wood fire is a great pleasure, almost everybody agrees. Unfortunately, a fire on the open hearth is also a dead loss when it comes to heating the house. The shallow Rumford fireplaces that graced each room of New England houses built in the early nineteenth century were reasonably efficient at throwing heat, but most modern fireplaces are not. But no matter how well designed, any fireplace in a house that has central heating draws a stream of prewarmed indoor air up the chimney. Fireplaces subtract heat from a modern house and add to fuel bills. This dulls their beauty, unless the artistic intention is to celebrate opulence.

The airtight box stoves that became popular during the 1970s were an esthetic bust. In the first place, the enjoyable qualities of an open fireplace were mostly shut up and lost from view. Worse, when the stove was crammed full and closed down for the night, all the complex hydrocarbons in the wood were volatilized into a sickly yellow smoke too cool to combust, before the residual charcoal slowly burned. Some of this tarry smoke recondensed as creosote in the flue, creating the hazard of a chimney fire—the rest went out to foul the neighborhood. Streets were lined with these little charcoal kilns, one of the most noxious traditional wood industries brought inadvertently back to life. At least our ancestors had the sense to make their charcoal out in the woods.

Woodstoves have since been redesigned, thanks to EPA regulations. Properly operated, the new stoves burn cleaner and yield more heat. They achieve nearly complete combustion either by a catalytic converter that lowers the ignition temperature of the exiting smoke or by directing streams of hot air over the top of the fire, igniting the gases within the stove. I like the second design, even though it is a shade less efficient, because it is more elegant and doesn't involve some finicky platinum doodad. Better yet, my stove feeds cleansing hot air along the inside of the glass door, so the soot burns away and the window stays brilliantly transparent. I can sit and watch this awesome secondary combustion taking place (which is about all I'm fit for when I come out of the woods, anyway). When a fresh log ignites it looks as though I have secretly opened a valve controlling a propane tank hidden beneath the hearth: jets of blue and yellow flame come shooting out of little holes in the back wall of the stove and spiral across the top of the fire directly toward the viewer. It isn't quite the same as an open fire, but it is definitely entrancing. Guests are suitably impressed. The chairs in our living room are seldom turned toward the television.

Now, I don't think very many suburbanites are going to heat their houses entirely with wood anytime soon, any more than I have been claiming we should grow all our own food in Weston. But these efficient new woodstoves take the dilemma out of woodheating. Esthetics and economics combine in them like oxygen and carbon. People can fire up the stove when they come home in the evening and enjoy all the cheer and company of a fire that they can see and still be contributing substantially to heating the house. They can use the stove as much or as little as suits their habits. When the day comes that oil prices rise again, people can adjust accordingly. If a thousand or so households in a small, semirural town burn a cord or two each for enjoyment and supplemental heat, taken from a few thousand acres of surrounding town forest, we are within the realm of sustainable local forest production.

The principle I have illustrated here with firewood obviously could be extended to our home furnishings and to the construction of the house itself. Every object that we take from the forest and bring into our homes to use and enjoy deepens our enjoyment of the living trees. There is beauty in the woods. There is beauty in a wood fire. But walking in a woods that is full of the smell of seasoning woodpiles and then watching that wood burn at home the following winter creates a powerful esthetic and functional bond between the home and the landscape. Each pleasure reinforces and completes the other, outside and inside. There is beauty in wildness, too, and in the wild things that share the forest and have no direct utility, I'm not denying that. In fact, our enjoyment of wildness will be justified only when it is not contradicted by the ecological destructiveness of our domesticity. We will see more of the beauty of the forest as a whole and take better care of it when we surround ourselves with local wood and live as forest people.

In the end, what Land's Sake does in Weston's forest is of little value if it isn't repeated in other towns, just as an exemplary shelterwood harvest in a single woodlot doesn't amount to much unless it is part of a townwide forest program. Good forestry can be carried on only in small pieces, but it becomes good forestry only if the pieces combine to cover large areas. This applies as strongly to the protection of biodiversity as it does to sustainable wood production. We certainly need Natural Heritage programs to preserve the habitats where endangered species live today. But this will do little good if these species do not have ample populations with ample room to move as disturbances come along and climate changes. What we need most is a continuous, permanent matrix of forestland across the landscape, covering all kinds of physical habitats, so that forest species can continue to flow with the ecological tides as they have always done. We need lots and lots of forest, an agricultural and residential landscape irrevocably bound by forest.[15]

How much forest is enough? We don't really know. I agree instinctively with conservation biologists who advise that about half of the landscape should remain in its natural state. I know from my own historical research that when forest cover dropped below one-quarter in this region there were dramatic ecological consequences. Half the land in forest seems a reasonable goal if we are to safeguard the normal ecological workings of the landscape. We need to protect this forest from excessive agricultural clearing, from excessively simplified management practices like large-scale clear-cutting and monoculture plantations, and from extravagant residential development. We got our forest back by sheer luck after screwing up once. We mustn't lose it again.

But how can we possibly ensure that enough forest will survive? Most of the forest in New England and throughout the Northeast is privately owned in small pieces and is essentially vulnerable. If these private forestlands were part of a prosperous, stable, small-scale farm and forest economy and were accorded their real value they might be safe in private hands, but in the present economy they are simply waiting for development. The average Massachusetts forest owner holds the property for less than ten years. If the current owner does not sell to a developer, the next one will—in the end the market will have its way. The prospect of state or national governments acquiring much more of this land for public forest is slim and not very attractive in any case because this would remove the land from local control. Large private conservation organizations such as the Nature Conservancy and the Massachusetts Audubon Society may be instrumental in preserving some key parcels that are under immediate threat, but is this an adequate basis for the widespread forest protection we need to see? We need to preserve not just uncommon, endangered forest ecosystems, but *common* forest.[16]

A true common forest will be made up of many local forests. The most workable, democratic alternative is for the forest to be protected at the community level, either by towns themselves or by locally controlled land trusts. The acquisition of open space has been under way in New England towns for the past few decades, but we need to raise our sights considerably. Only about 6 percent of the forestland in Massachusetts is owned by municipalities and private nonprofit organizations. Towns like Weston and Lincoln have protected nearly a quarter of their land from development, which is better but still not good enough. We should aim for *half the landscape* in locally controlled common forests.

Every town in New England and beyond should create a town forest of at least a thousand acres, ranging up to ten thousand acres or half of the town. This forest may be composed of a mixture of community-owned land and private land covered by conservation easements. The primary purpose of this common forest would be to protect the ecological integrity of the landscape and to safeguard natural biological diversity—but that is not enough. Environmentalists are already sold on the necessity of such ecological protection, but we must recognize that this motive alone is unlikely to galvanize the political support necessary to create forested commons on an adequate scale.

The wilderness ideal, standing alone, is fatally flawed as an approach to saving the forest—it is historically unfounded, ecologically marginal to the world as it now exists, and elitist. Of far greater ecological value than a few hundred acres set aside as wilderness would be a few thousand acres of productively managed, quasi-natural forest in every town. Blocks of wildland at every scale may be appropriate, and it might be wise eventually to designate a chain of very large wild areas down the Appalachian spine of the eastern forest, as some wildlands advocates are proposing. I would have no problem devoting a substantial part of national forestlands to a wilderness core, if I knew that the great bulk of the surrounding forest were securely held for sustainable forestry, thereby providing work for rural people, renewable resources, and long-term ecological integrity on a scale not conceivable through static wilderness areas alone. To me, this means steadily gathering private land into thousands of locally controlled commons. The key to creating and maintaining common forest—indeed, perhaps the only force that can generate enough passion to protect sufficient forest—is active engagement of people with the trees.

Involving people with trees has traditionally meant providing trails and recreational opportunities and perhaps using the forest as an educational laboratory for schools and scouts. More recently, in some places it has come to mean ecological restoration projects that inspire volunteers to plant seedlings of threatened native species and combat exotic invaders. But to work best, involvement must also mean real economic use, revolving around regional consumption of high-quality forest products. Forest protectors must form a coalition with forest workers and make a different kind of wood products industry in the process—one that at every step emphasizes quality work rather than lowest cost. The harvesting and marketing of wood has to be done with restraint, in ways that make people feel proud to live as part of the forest. Russell Barnes once remarked to me that a thousand acres of New England forest properly managed could keep a small logging company busy cutting wood forever, and our experience at Land's Sake bears him out. There are many satisfying forest jobs waiting to be created in this region, if environmentalists spend their money where they live and buy locally produced, high-quality wood products. Private logging contractors will have to meet the exacting standards that we as common forest owners will insist on and that private landowners are also coming increasingly to demand. One nonprofit community forestry company like Land's Sake in every town would serve to train loggers in this kind of work, sending a steady stream of young people to the fine forestry programs at colleges in the region. Some of them would return to manage their hometown forests, where they would teach the next generation the joys of splitting wood. This is not a pipe dream—I have seen it work.

Many of New England's original towns began with the great bulk of their forestland in commons. New Englanders have been unconsciously reviving this tradition over the past few decades through their town conservation lands because it makes sense. The forest bestows common benefits: clean air and water, biological diversity, and beauty. These benefits can be realized only when a sufficiently large part of the landscape remains in forest—a piecemeal forest is worth much less. These values are not marketable by private forest owners, however, and hence exist precariously on private lands only as a consequence of forbearance or inadvertence. The forest is the part of the landscape that ought to become commons again. I do not object if private owners who know and love their woodlots retain part of the forest in private ownership, but I have no confidence in them in the long run—not in today's unbridled market economy. A sufficient core of forest needs to be held in common. It should be acquired by towns and by groups of citizens acting through the market, and it should be managed democratically at the local level. The forest is the ecological key to the New England landscape. It gives the region its character. It holds our common heritage and our common future, and we ought to hold it as commons.

Spring at the farm. (Neil Baumgarten)

Chapter 6

Reclaiming the Commons

One Sunday morning in April 1997, Faith and I took our first walk in a few years around the Case Farm. I had not worked at Land's Sake (except as an occasional volunteer) since the spring of 1992. For several happy years we lived in Salina, Kansas, where I worked for the Land Institute while Faith delved and span, wove and planted. We returned to Weston only for brief winter visits to see family and friends and to get out in the woods on skis. Kansas is a nice place, but it is a bit short on trees and snow—just as New England is short on grass and sky. Now we were back in Weston for a good while. To be at the farm on a fine spring morning was to be home. With us were Jake, the retired sheepdog, and a baby boy named Liam sleeping in a sling beneath his mother's shirt.

The farm looked good. Land's Sake is prospering these days; a shiny company dump truck is parked by the stand. The earth had just absorbed a colossal April Fool's blizzard that dropped nearly three feet of snow and brought down more treetops and branches than any storm since the night in

May 1977 when we voted through the purchase of the Weston College land at town meeting. Damage at the farm was minimal, fortunately. The magnolias were already blooming and the cherries were about to. Most of the fields had been disked for planting and the hay mulch pulled from the strawberry beds. The summer raspberries were neatly pruned, and the new canes were just breaking ground. Rhubarb was poking up in its bed around one of the Norway maples. Ed Hagenstein, who succeeded me as director and then moved on to other things, was nevertheless out digging beds in the perennial flower garden, his pet project. Tree swallows dipped over the fields in the bright spring sunshine, and a pair of bluebirds defended one of the nesting boxes. The towering Norway spruce by the parking lot died long ago, but the little spruce we planted beside it was flourishing. It is hard to describe how elated Faith and I felt to see the farm growing beyond us, in the care of other loving hands.

Now that Land's Sake has a history of its own, has become an established part of Weston's character, and seems to have a rosy future, I can see it as part of an ongoing story. What we are doing grows from what was done before us. We did not just make this up on our own, although it seemed that way at the time. I had not quite remembered how vividly the farm's beauty derives from the interplay of the small cultivated fields with the open parkland and tree plantings, a landscape we inherited. Surrounding the fields are the noble survivors of the dozens of trees planted when the textile merchant James Case turned an old farm into an estate more than a century ago—the Norway maples and spruces and the grand European beech trees on the hill. Beneath them is a layer contributed by the Arnold Arboretum half a century ago, the flowering crabs and cherries, silver lime and mountain ash, sourwood, silverbells, and dawn redwoods. The town and Land's Sake carry on not only this esthetic tradition, but also a tradition of educational farming for children established by James Case's daughter Marian at the beginning of the century.

When we go to work in a community we make ourselves responsible for things we may not even have been aware of when we started. Working with Bill McElwain at Green Power and with Doug Henderson and Martha

Gogel to form Land's Sake, I aimed to help create something new in Weston—a model of organic farming and local self-reliance that was to be the harbinger of the ecological age we expected momentarily in the 1970s. I found instead that through our community farm we became the keepers of a small flame that was lit before any of us was born. Our duty is to keep that fire burning in the midst of a hurricane of suburbanization that has blown much longer and has swept much farther inland than we ever dreamed it could.

Marian Roby Case founded Hillcrest Gardens in 1909 on land she had acquired adjacent to her father's estate. She was forty-five years old at the time, which I find encouraging. Her gardens and summer school flourished through 1942, an inspiring run of thirty-four seasons. At its height, Hillcrest Gardens grew forty acres of orchard fruits, grapes, small fruits, vegetables, and flowers by the latest methods of the day, selling the produce in Weston and surrounding towns. Every summer, ten to twenty teenage boys mostly from Weston were provided with jobs and with an education in horticulture and the natural sciences. There were lectures and field trips, uniforms, prizes, and songs. The gardens ran at a loss, and Miss Case subsidized the operation for its entire existence. She was an aristocrat who found her life's work instructing the sons of Weston's working class in gardening and instilling in them an appreciation of nature, "fulfilling the old-time spirit of noblesse oblige," as she put it. Many of her students did indeed go on to careers working the land, from nurserymen to planners. It is interesting that she took up her task just at the moment when market gardening as a practical enterprise had passed its zenith in Massachusetts and was going into steep decline.

I knew little about Hillcrest Gardens when Land's Sake was founded and by seeming coincidence put in place a remarkably similar program on another part of the Case property. But perhaps it isn't so surprising. I cannot believe it was *just* a coincidence that we followed so closely in Marian Case's footsteps. Later I learned that Doug Henderson's brother had been a Hillcrest boy. Doug and others in Weston did indeed remember what Miss Case had done, partly with admiration and partly with affectionate humor, because she was apparently a bit of an "odd stick," as someone once remarked. The disappearance of Hillcrest Gardens left a hollow in the community that resonated with the call Land's Sake later sounded. It seems only fitting that a program such as ours should reappear on that land, like a mushroom after a rain that breaks a long drought.

The history of a community is full of such echoing events. The things that give a place its peculiar character, good and bad, are carried down through generations in recurring dramas of whose antecedents the actors themselves are sometimes only dimly aware. Good examples such as Hillcrest Gardens are seldom lost—they have a way of resurfacing in new forms. I keep this in mind because I have learned that changing the way we treat land in this country is going to require more than one generation—it already has. Many fine efforts will flourish for a time and then disappear, only seemingly to be forgotten. Land's Sake may one day do the same, but it will not have been in vain. Not only will we have carried on Miss Case's legacy, we will have been part of bringing back into community consciousness a much deeper memory of the value of common land, an idea hidden within the woodlands that have themselves recovered from near oblivion.

I feel a kinship with Marian Case, even though I am uncomfortable with her patrician philosophy. She would no doubt feel the same way about me, with my talk of commons. Still, Hillcrest is special to me because of its attempt to combine farming with education and because of the large number of young people it inspired. Land's Sake has now been farming the Case land for about half as many years as Miss Case ran her gardens. Between us, we account for more than half of the twentieth century. Between us, we have seen to it that a wonderful stretch of country near the center of Weston has remained strikingly beautiful and has played a central role in this town.

This land was not intrinsically more important than any other land in Weston, but it has become so. It has become something to live up to. When development threatened—as it inevitably will threaten *any* privately held land because everything in America is for sale—the town rallied and saved it. When people come house-hunting in Weston today, the first place the real estate agent takes them is to Land's Sake—to show off the protected rural character of the town! Now the latest real estate boom engulfs us, and we struggle to protect the last remaining open spaces before they too are crowned with houses so grand that even Westonites' jaws drop. As the same storm breaks over towns thirty and forty miles from Boston, it is time to examine the larger significance of the small victory we have won in Weston, with our common land and community farm. What good does it do?

The suburbs are coming. In many places the suburbs are here, and in many more they are swarming just over the horizon, their vanguard leaping ahead along the ridgetops to command the best views. I was born in suburbia, and I now fully expect to be buried in it. I am not happy about this, but every generation has to meet its fate. Americans are going to continue moving to rural places and small towns. Unless we adopt better ideas about land ownership and care, Americans are going to continue ruining these places and bankrupting the future. Suburbanization is an impulse that destroys the very rural character it seeks. Therefore, those of us who think we have some better ideas about how rural land might be reinhabited must consider how we can harness the suburban impulse and turn it toward a better end.

Suburbia, which epitomizes and requires the profligate consumption of nature driven by industrial capitalism, is probably not a sustainable form of settlement. I certainly hope not. In this book I have summarized the reasons I believe it must be stopped, or rather transformed: the necessity to reduce our rate of resource extraction and excretion to what the biosphere can tolerate, the necessity to grow food and wood in ways that protect our water and soil, the necessity to safeguard the biological diversity, ecological health, and beauty of the world that enfolds us. Understood in these terms, a thriving rural world is necessary to our survival, but we are currently treating what is left of that world as a fleeting luxury to be consumed by those who can race momentarily ahead of the rest. This obviously can't go on forever. Unfortunately, the suburban drive no doubt will be sustained for several decades at least—long enough to do much more damage. In one way or another, suburbanization will occupy and change many more rural places. We need to see to it that rural resettlement takes a new form that permits and even promotes better ways of living.

Champions of small farming like Marty Strange, Gene Logsdon, Wes Jackson, and Wendell Berry have long warned that we are losing our family farm culture and rural communities. For generations now, the bell has been tolling for agrarian America. Where the agrarian tradition still lives it is worth fighting to defend it. Deep in the American heartland are regions that continue to lose population, that desperately need the return of people to take better care of the land and make depleted communities whole again. However, it is now clear that the problem confronting many other rural places is not simply how to prevent decline of one kind, but how to survive growth of another. Carried along by cheap transportation and ubiquitous telecommunications, the affluent can now live wherever they choose. Having laid waste to the countryside within commuting range of the beltways, they covet second homes and vacation chalets in more remote places of particular charm. No hilltop in America is safe from them, as they seek refuge from themselves. The rural diaspora we agrarians have long dreamed of is taking place before our eyes, in the worst way we could ever have imagined. We have lived to witness rampant decentralization of consumption, paradoxically driven by continued centralization in the control of productive land and the extraction of natural resources.[1]

This residential flood drowns local places, rather than nurturing them. Conflicts inevitably arise between those few who were born in rural areas and make their living by working the land with efficient industrial machinery and the nouveaux rustiques who understandably would rather not live next door to such enterprise, even though it ultimately feeds and houses them. Ironically, many of these rural newcomers consider themselves environmentalists, but their ideas about what constitutes a healthy relation with the land tend to be confused at best. Their often passionate desire to leave land near their homes in an imagined pristine state of nature is in stark contradiction to their voracious consumption of natural resources. This is duplicity in place of atonement.

Our society rewards naked exploitation of nature and indulges naïve, romantic love of nature, but the middle ground of caring for nature while using it productively has been declared uninhabitable. Our economy drives us to use nature as cheaply and unattractively as possible, and to compensate we pay extravagantly to enshrine natural beauty in a few selected unproductive parts of the landscape, ranging from showpiece yards to wild refuges. The wild areas cost us nothing but a little lost production; landscaping suburbia consumes enormous expenditures. There is an odd dualism, or even trebleism, at work in our attitude toward land here. We demand that land be either cheaply used, expensively manicured, or utterly untrammeled in order to fulfill separate functions. Any suggestion that we might farm and log in restrained, beautiful ways in the first place is dismissed as inefficient and nostalgic at best and a sentence to misery and starvation at worst. This is economically absurd, but that is our economy.

We need a new attitude toward the land that sustains us, what Aldo Leopold called a "land ethic" half a century ago. But how do we get it? Can we engender a new, broad-based agrarianism that embraces both rural natives and suburban newcomers, that both protects and uses the land, under such polarized circumstances? I believe that we can, and that such a *common agrarianism* is urgently required. It will be by nature a bit less efficient at extracting resources from the land but will provide other benefits that ought to have economic standing. The keys to it are, first, reclaiming our common interest in the land; and, second, creating local economies that fit the land well and that actively engage people who move into and grow up in rural communities. We need to develop a common culture of caring for land, and we need more common land under our care.

Here are two plied but distinguishable strands: the way we own land and the way we use land. I want to address first how to better balance private and common interests in owning land, and then how to build strong local economies and culture. This is a call to the commons. But in calling for more common land I am *not* advocating a sweeping collectivization of agricultural production. Commons systems are never founded on exclusively communal ownership of property, and state systems that have approached this extreme have been clumsy, brutal, and ugly. I am not proposing to turn the suburbs into agricultural communes, but I am tired of watching private greed trample common good at every opportunity. Neither am I proposing an atavistic return to medieval common field farming and all that went with that, from the oxcart to the Black Death. Living in such a world would make me sick—

in fact, I would already be dead several times over. I am equally queasy, however, about the fate of soil, water, food, and forests under the free market colossus that has been advancing for five hundred years and that now confidently bestrides the globe. It is clear to me that exclusively private ownership of land and extraction of commodities in a market economy is a better-paved road to a bigger ruin. We need a mix of private and common interest in land that is appropriate to the world today, one that balances personal freedom and community responsibility, economic efficiency and ecological restraint. In my view, this means more commons; but it also means a strengthened role for small private owners and entrepreneurs, as opposed to large corporations—which call themselves private even though they are plainly oligarchic collectives. Corporations that are devoted to maximizing profit have no rightful business owning land, which needs to be cared for in ways that necessarily reduce profit. The call to the commons must also be a call to the private rights of commoners.

Where does the common interest in land lie? That depends on the place. The balance between private and common interests will obviously vary from one kind of land to another within a community, and from one community to the next. Some kinds of land are best held privately, some commonly, and some with a mixture of private and common rights. We can think of the common interest as ranging from regulations that limit the way land may be used, through easements that convey restrictions permanently to the community for safekeeping, to outright common ownership. In general, the more inhabited and intensively cultivated the land, the more it belongs in private hands, and the more uninhabited and lightly managed, the more it belongs in commons. That is, as we move from the household toward the wild, we should also move from the most private toward the most common.[2]

Each community needs to determine an appropriate balance among these elements—there is no perfect formula. But we can do far better than we have been doing. Weston now has coming on one-quarter of its land in commons and most of the rest in houselots, which is a great achievement given the town's proximity to Boston and the era during which it suburbanized. Towns such as ours that have endured half a century of intense, often white-hot development pressure are the crucibles in which new land values have been forged. That land may have common as well as private value is still an awkward young notion in our time, but at least it has been reborn. Other rural places must learn to embrace and nurture this value earlier in the struggle, to protect more land. What balance of ownership might be achieved in a town that is just beginning to suburbanize? For simplicity's sake, let me examine three categories that stand for the full spectrum of land uses: residential land, farmland, and forest.

First, residences. I have no quarrel with the American ideal of private home ownership. It is entirely appropriate that a community contain a substantial number of single-family dwellings. (It is also appropriate that a community prevent the rising value of land from pricing out the less affluent. A community needs to make home ownership affordable for most of its citizens and also to have other kinds of decent housing available. Weston has failed this responsibility.) I *do* have a quarrel with the extravagant five- and ten-acre lots that often now surround houses being built on the suburban fringe and with the far-flung dispersal that results. A more corrosive pattern of settlement could hardly be imagined. Such detached housing, as it is aptly termed, ends by expressing extreme individualism, gluttony, and ostentation. Trophy houses, we call them here. For a family to enjoy private space for a yard and gardens is fine. But as houselots grow in size and multiply across the landscape, farmland and forestland are fragmented and the commonly enjoyed rural character of the place is rapidly consumed. The common interest should be to keep residential (and commercial) development confined to the least possible space that will accommodate a healthy rural community of a few thousand people—let's say not more than one-quarter of the land area and preferably far less.

Reclaiming the commons requires a new vision of the benefits and responsibilities of private home ownership within a community, a vision that extends beyond the garden fence. It will require a dramatic mental shift for upwardly mobile Americans to be satisfied with close quarters in place of the spacious suburban lot. People do not give up their dream of detachment easily. Those moving into rural communities must come to believe that it is in their best interest not to occupy as large a piece of land as they can, but rather to occupy as small a parcel as they can in order to enjoy more neighboring farmland and forest instead. Each new building lot should be counterbalanced by a deposit of land into the commons. Rural newcomers must accept their responsibility to protect the pastoral character that attracted them in the first place and to actively support a strong local land-based economy. That is, with a small parcel to call their own, nonfarming residents need to recognize that living in a community with attractive working farms and access to common forests confers private benefits that outweigh engrossing a large, expensive miniestate of their own; and that they can't have both. For better or worse, the era of James Case and others like him—the wealthy elite that could afford large estates in rural towns close to cities—ended half a century ago. Today, a much larger, more mobile affluent class subdivides and destroys. But why divide a one-hundred-acre farm into five-acre farmlets when the same twenty houses and their gardens could be confined to one corner of the property, leaving the farm largely intact and functioning and in some measure open for all to enjoy? I will have more to say about nurturing such nascent agrarian aspirations among an increasingly residential rural population in a moment.

Farmland lies in the middle of the spectrum of private and common interests in land. I am all for private family farms. I am convinced that in time the economic advantage will return to those who work smaller pieces of land in more intensive, diversified, sustainable ways than the average commercial farm of today. I hold this view because I believe the cost of industrial inputs will rise, tolerance for environmental damage will run short, and demand for fresh, wholesome, locally grown food will continue to grow. Working the land intensively will require native farmers who know their place well and care for it with devotion, a breed we have seen far too little of in our history of occupation to date. Experience has taught me how long it takes to get to know land and climate, how long it takes to integrate all the elements of a diversified farm. Small farmers can best build such a culture of competent local understanding and affection, as Wendell Berry has long argued. Given an even break, small private farmers have stronger incentives than either corporate or collective farms to work the land efficiently and to get the most from it—this has been abundantly proven. Given an ecologically sound economic framework and a secure footing, they will take good care of the land. This is why we need a tenacious private hold on most farms, preferably a hold that lasts for generations. The question is, How can we keep such people on the land?

We must turn to our common interest in farmland. Obviously, the common interest is not being well served by most private landowners today (including most farmers) because the present economic climate does not encourage deeply rooted small farms. Far from it: the present climate promotes unsustainable farming and the prompt sale of farmland to a higher and better use—gaudy palaces and tacky marts. This situation is likely to continue for some time. So how can the common interest in protecting farmland best be served today? Private farmers cannot meet it on their own—we need to help them along by imposing reasonable restraints. We must insist (as we increasingly do) that farming not cause ecological damage, such as polluting our water or allowing our soil to slip away at irreplaceable rates. We cannot legislate the best farming, but we can at least outlaw the worst. We can also subsidize the best practices, such as those that protect our watersheds. Such measures may increase our taxes or the price of food at the cash register, but that is as it should be. It will begin to swing the competitive advantage toward those farmers who best care for the land, where it belongs. Of course, this assumes that subsidies and regulations as they are actually written and administered can really promote good farming and the public interest in healthy food and land, instead of the interests of those best able to manipulate or evade the laws—a rather large assumption. But the alternative is to grant those same powerful interests a free market to work their will with the land, which is far worse.

Our strongest common interest lies in simply keeping farmland from being lost to development. We need to decide as particular communities that we are going to *need* local farmland in the not-too-distant future and that we *want* it even now, when the future need for it is still a matter of debate. If our goal is to keep family farmers on the land, preserving the possibility of the best farming, then our best tool is the conservation easement. By an easement, the community sequesters the right to develop the land, while the private owner remains otherwise in possession. In this way, both private and common interests are served. The farm family stays on the land but is taxed only on the residential value of its houselot (like any other family) and on the agricultural value of its farmland and improvements (like any other business). The exorbitant potential value of the farmland for residential development is absorbed by the community as a whole, thus expressing the common interest in preserving attractive, productive agricultural land close by. Either land trusts or municipalities can acquire and hold such easements. The difficulty, of course, is raising enough money to protect more than a token amount of the farmland within a community and doing so equitably. Our goal should not be modest. Our goal should be to keep anywhere from a quarter to half of the landscape (or even more depending on the terrain) in agricultural production—a monumental challenge at the urban cutting edge.

Clearly, the degree to which the common interest in protecting farmland needs to be actively asserted and the cost to the community of asserting that interest will vary greatly depending on the amount of development pressure. The need for common rights in farmland runs from slight in the hinterlands to severe at the city limit. Far away from the expanding metropolises, in happy regions that do not enjoy the mixed blessing of postcard scenic charm (but are only commonly beautiful), there may be little need to acquire conservation easements. Where demand for country estates isn't inflating the value of land, it will remain open on its own. At the other extreme, where very little open space remains there is much to be said for acquiring what is left for outright commons. Where people are thick there is an overwhelming need to secure common land for public enjoyment, whether as community farms and gardens or as parks. In between these poles lies the countryside where there is a common interest in keeping private farmers on land that the market now reckons is more valuable for dispersed housing. The need for easements is strongest in places that are suburbanizing or enduring an influx of vacationers, snowbirds, or some other breed of wall-eyed urban refugee. The earlier the need is recognized, the more we can accomplish.

At the far end of the spectrum of intensity of land use lies wilderness, and close beside wilderness lies sustainably managed forest. Forestland belongs largely in common ownership. Ecological integrity and biodiversity provide little profit to any individual proprietor. Sharp accountants consider them economic liabilities. Therefore private owners have scant incentive to protect these values, beyond enlightened affection—which is admirable but doesn't last long in the market. Eventually, individuals who love the land for its own sake are forced to sell, can't resist selling, or die and leave the land to someone less affectionate. The *common* interest in ecological health is very strong, however. In the long run it is crucial that the ecological backbone of every community be permanently protected and placed under integrated local management. This might include several large, connected stretches of forest (or wetlands, savanna, prairie, as the case may be), amounting to several thousand acres in every town. Ideally, the community itself should own such conservation land and manage it for biodiversity, passive recreation, and sustainable wood and timber production.

Locally controlled common forests are desirable even in sparsely inhabited communities that are under little immediate threat of development. Common forests offer a secure basis for management at the landscape scale and place the means for local livelihood under local control. Privately owned forests are always vulnerable to acquisition and exploitation by outside interests, while state and federally owned public forests are too often effectively controlled by the same well-financed interlopers. By creating community forests we may also build the political constituency necessary to force better care of adjacent state and national forests. Locally owned common forests are certainly not immune from abuse, but I believe they provide the safest, most democratic building blocks for the necessary continuous matrix of healthy natural ecosystems across the continent.

This is not to say that all forestland needs to be commonly owned—there is plenty of forest out there for farmers and others who wish to own private woodlots. But owners come and owners go, while the urge to ravage forestland with either a feller-buncher or a bulldozer never sleeps, except with one eye open. Let us aim for an ecologically adequate chunk of permanently protected common forest in each community. Let private woodlots attached to surrounding farms function as more flexible adjuncts that can move from forest to agriculture and back over decades and centuries, as the market and landowners determine. As a practical matter, of course, the forest of most communities will probably be made up largely of private forestland at best protected by easements, rather than outright commons, for a long time to come. That is a good place to start. Eventually, at least one-quarter and ideally one-half of the land in every rural community should lie in common forest and other natural ecosystems appropriate to the region.

There is another aspect of the landscape that deserves common protection as much as its biological health: the most prominent hilltops, the shorelines of rivers and lakes, and other places of transcendent beauty. These also ought to be commonly owned, and no one ought to squat there, as Henry Thoreau insisted long ago. "Think of a mountain top in the township . . . only accessible through private grounds," he wrote. "A temple as it were which you cannot enter without trespassing—nay the temple itself private property and standing in a man's cow yard—for such is commonly the case. . . . That area should be left unappropriated for modesty and reverence's sake—if only to suggest that the traveller who climbs thither in a degree rises above himself, as well as his native valley, and leaves some of his grovelling habits behind."[3]

Yes, think of it. In Thoreau's time the trouble was that the private owner did not appreciate the beauty of the temple but saw it with the eye of mere utility. In our time the owner appreciates the beauty only too well but wants to engross it for himself—and not even be bothered to go outdoors and climb for it, but to guzzle it without rising from his easy chair. So much for modesty and reverence.

Like so many private appropriations of the commons, this self-aggrandizing impulse toward scenery is self-defeating in aggregate. What commands the view also occupies the view, and soon enough every noble outlook is defaced, and every private vista commands only similar eyesores. The picture window becomes an unerring mirror, reflecting inward. For the rest of us down below, a broad underview of all our superiors' groveling habits is fully revealed. Perhaps a single castle on a hill acquires a certain rustic charm after a thousand years or so, but when three and four and then a few dozen pop up, all the natural contours of the landscape are welted and scarred. People ought to have the common sense to build lower on the slopes, but of course as detached individuals they don't, vying instead for the peaks. Protecting the charming face of a community is just as important as protecting its ecological bones, and fortunately the strongest skeleton often underlies the most handsome features. Those working to build common forests should secure the scenic ridgetops and riparian areas first, before the speculators ride into the valley with their fat wallets and narrow vision, scanning the horizon for prospects. The best lookouts in Weston belong to our commons, and we can all climb thither in peace, without glancing uneasily over our shoulder for the puffed-up proprietor of the day.

That is a vision of how rural communities might be reinhabited and kept whole. With a little foresight, development could be confined to a small percentage of the land in a central village and outlying hamlets and clusters, while a mixture of private farms protected by easements and forests owned in common could possess the balance of the landscape in perpetuity. Needless to say, in the real world acquiring such easements and commons will be painfully costly and slow. We will inevitably fall far short of the ideal and for a long time lose much more than we save. In many places it is already too late to achieve anything like an ideal resettlement, but we must keep working to do what we can along these lines—every small victory gives us more to work with for the future, more land with which to change minds. Someday, resettling rural areas in such a way may become a necessity, and the more we know about how to do it, the better. We must be like the beleaguered forest a century ago, holding on to all we can and sending forth our ideas like seeds, encased in enough accumulated experience to nurture them. A few will fall on fertile ground.

At the moment, we are in an unequal race against unbridled development that is rapidly harrowing the landscape, with seemingly endless ready cash to spend. One may well wonder where the countervailing funds for commons will be found. In some cases public-spirited landowners may donate easements for love because they wish to ensure their land remains undeveloped and to reduce their property and inheritance taxes. But more often common lands and easements will have to be purchased by the community. Local citizens will have to be democratically convinced that it is in their interest to tax themselves for this purpose, as has happened in Weston. This may save a few key parcels of land, but that will only get us so far.

Fortunately, the force of invading development can sometimes be turned against itself—the art of self-defense judo for threatened land, if you like. One approach is to impose a tax of 1 or 2 percent on every sale of real estate, the proceeds going into a land bank to purchase conservation land. This is really just a property tax in a palatable and appropriate form to address a specific need. In this way a small part of the rising price that newcomers are willing to pay for property can be used to protect open space for public use. Maryland and Vermont have enacted such taxes. Massachusetts has successful land banks on Martha's Vineyard and Nantucket, and now a wave of towns are pushing for such measures in spite of stiff opposition from the real estate lobby. By floating large bonds against an assured source of revenue stretching into the future, communities can protect a substantial amount of land before it is too late.

Another way to limit sprawl is to promote partial, clustered development, in place of cookie-cutter build-out. In this way part of a threatened parcel can be densely developed, thus protecting the remainder. Clustered development (or conservation subdivision, as it is sometimes called) appears to be the most effective tool we have to save land, and town planning boards should be going all out to encourage it. Sometimes complex deals can be put together that include the clustering of a reduced number of homes, a conservation easement over most of the parcel, and the outright purchase of the most attractive part of the land for commons. Clustered developments can be ecologically well designed themselves, and well integrated with the protected open space surrounding them.[4]

All this marks a radical departure in how we own land, no matter how it is couched. Transferring anything like one-half of the forest from private hands to commons and holding protective easements over a substantial portion of the remaining private farmland would of course mark a fundamental transformation of American landownership. It surely will not happen overnight. But if it does not happen over time it is hard to envision how we will be able to protect the farmland and forest we need to live sustainably, let alone enjoyably. A revolution in landownership is needed. But I want to repeat that I am by no means calling for the forced expropriation of private property, for a vanguard of the eco-tariat to impose collective farms, cattle-prod in one hand and Kalashnikov in the other. I am not demanding that corporate middle-managers surrender their backyards, shoulder their grub-hoes, and march to the common fields for reeducation. I envision towns in which private and common property rights flourish side by side and often overlap, drawing on the strengths of each. This is in fact what we are starting to see in towns such as Weston—hardly a hotbed of socialism.

Thoughtful agrarians are only selectively nostalgic. The point is not to return to the Middle Ages but to adapt the principle of the commons to the modern world. In the medieval world of resource scarcity, traditional systems brought together common and private control over different kinds of land in useful ways. Commons systems were designed not to maximize production but to optimize ecological security. This was a conservative approach to landownership, and I think we will need something like it again as we approach a new set of ecological limits in the world. Lop off the overlords, insist on protecting ecological values on common forestland, acquire a measure of common protection for private farmland, and we will have a very effective and appropriate mix of landownership for the coming millennium.

But aren't commons by their very nature hopelessly inefficient, if not prone to tragic abuse? I need to say a word here about the widely held but mistaken belief that commons are invariably wrecked by the commoners themselves, whereas private owners tend to take good care of their land—the so-called "tragedy of the commons," in Garrett Hardin's unfortunate and misleading phrase. Hardin coined the term as a metaphor for the unlimited right to bear children in a crowded world, but it has been heartily endorsed in its literal sense by the property rights movement. The theory holds that because an individual has a strong personal incentive to cheat on the commons and a much weaker incentive to care for it, any commonly owned property is bound to be overexploited and thus degraded. This profoundly underestimates human nature—not private virtue, but collective jealousy. In actual historical fact, virtually all commons were closely guarded against individuals' cheating by the watchful eyes of their neighbors, backed up, when necessary, by the law. Commons were also vigorously defended against outside intruders. The term *commons* includes not only a common resource, but a mechanism for community control. There is no such thing as a free commons: if it's free, then it's not a commons. Commons have bounds and rules that limit access. Commons, like democracy, must always be defended. Given that, the system worked remarkably well for long periods of time in many parts of the world.[5]

The corollary of the tragedy of the commons argument holds that private ownership ensures better treatment of land because the owner has a strong incentive to care for it. This assertion is so childishly simplistic as to be disingenuous. At best, it profoundly overestimates private virtue and resistance to the temptations of the market. A powerful incentive to care for the homestead may exist where a family hopes to live in one place for generations, and that is why I favor private ownership of farmland—once the urge to liquidate the ancestral land for profit has been eased. But in most cases in our culture, private ownership is brief. It is simply a license to exploit the land for all it is worth and then to cast it aside while reinvesting the profits elsewhere. The owner is not obliged to live on the property after finishing with it and often never lives there at all. To understand that private ownership is no panacea, one need look only at the cut-and-run practices of nineteenth-century timber barons (or at the similar behavior of some corporate forest owners today, for that matter), who let the land become a public charge after they had stripped it. Private ownership certainly does have a productive role to play if limitations can be placed on the rights of ownership and if the market system in which it is embedded can be modified to give the owner ecologically responsible signals.[6]

Ideological quarrels that pit common against private property as ideal types are pointless. Each has its place. Common ownership cannot guarantee good management, but it is a sensible choice for that part of the landscape where the highest priority is restraint, rather than productivity. Private ownership is generally more productive but ecologically more risky. Even in long-settled New England, with its strong heritage of devotion to place, privatization of common forest and grazing land led in time to devastated woodlands and rundown pastures. Once the market system was firmly in the saddle by the nineteenth century, private farmers (some of whom represented the proverbial seventh generation) often chose to cash in their ancestral legacy by "skinning the land" rather than husband it any longer. Common ownership may not often lead to the most efficient production of wealth, but it is an appropriate means to safeguard long-term ecological health. The tragedy of the commons seldom lies in common ownership itself. The tragedy usually lies in the expropriation of common resources for unrestricted private gain. Privatization is seldom the solution to the tragedy of the commons—more often privatization *is* the tragedy of the commons. The solution is to reclaim and strengthen the commons.

The commons is not a utopian ideal of ownership of everything by everybody. There are things like the Earth's ozone layer that can be considered a global commons and need to be protected by international agreements. Our national and state forests are sometimes called commons, but, being so large, they must be controlled by bureaucracies; they become battlegrounds of competing interests. The kinds of commons I am discussing here are small, and the bounds are not far from home. They can be perambulated in a day. One of the crucial ideas in a commons system is that the land be controlled by the *local* community, by people who know it well and must live with the consequences of their actions upon it and with each other. This requires a community whose residents know how to speak to each other—which is perhaps why commons have reappeared in New England, with its town meeting form of direct democracy.

A commons also requires commoners who are productively engaged with the land, beyond weekend birding or mountain biking (I speak as an avid birder and biker, of course). For this reason a commons system and a revived local economy must go hand in hand. Such an economy will require broad participation, if that part of the community's land which is common is to be governed by residents who fully appreciate its value. We are not likely, however, to see the rise of a new agrarian class of Jeffersonian small family farmers any time soon, even though movement in that direction would be healthy and desirable. One can still sanely dream of such a revival taking place someday in parts of the Midwest that have been emptied, but on the densely settled eastern seaboard and in other suburbanizing regions that would be a flight of purest fancy. What we can envision here, I believe, is a broader, common agrarianism in every community that includes a few dozen full-time farmers and woodsworkers, a wider range of part-time farmers, and one community farming and forestry program that involves most residents with the land in some meaningful way. In this way, we can make caring for the land a normal part of growing up in rural, suburban, and even urban communities—a common experience, if you will. This will provide our commons with the necessary commoners.

To have a healthy relation with the land, we need places that are held together by diversified local economies. This is true whether industrial society is entering an era of continued prosperity and expansion or one of tightening limits, scarcity, and decline—or one gradually being overtaken by the other, as now seems most likely to me. In the coming century, our species will face enormous ecological challenges. We must make the transition to an energy system that no longer relies heavily on fossil fuel, meanwhile feeding, clothing, and housing some ten billion people without exhausting and destroying our soil and water. Next, we must do this without woefully diminishing the complex biosphere composed of millions of fellow-species that makes our planet habitable and beautiful. How are we to meet these human and biological imperatives at the same time, from the same ground? The population of the United States is expected to increase 50 percent by the middle of the next century to nearly four hundred million, and we are people who are in the habit of consuming exorbitant quantities of resources. Clearly, some changes in the way we use land are in store.

I am unable to predict how well we will do at meeting these challenges. In either case, whether we manage to spread material prosperity and comforts widely among the members of our species or suffer such setbacks that most of us barely scrape by at the level of survival, we will benefit from a new, ecologically sound, common agrarianism. It may be forced upon us by necessity or we may find it to be a healthy and rewarding way for prosperous people to live. The sooner we come to it by choice rather than dire need, the better.

Given the prospect of a relentless increase in demand for resources, it is likely that we will soon be looking to produce all we can from the land right around us, even in stony New England. But what can we best produce here, in an ecologically sound manner? The history of the region outlined in the previous chapters gives some idea of what products New England might concentrate on. It is important not to let our imaginations become bound by what is profitable in the present economy, skewed as it is by cheap fossil fuel and surplus production—even though in practice we have to deal with that reality for now. On the other hand, advocates of local economies simply make themselves look silly by insisting that every region can or should provide all its provender from local sources. Every region needs to consider what it could reasonably supply for itself, what it would still need to import, and what it could sustainably export in a world of closer ecological limits.

First, New England ought to be able to provide the great bulk of its vegetables and fruits from its own soil—certainly within rural towns and maybe for the urban population as well. Even the dense state of Massachusetts can do this with a small percentage of its land area. This can be achieved through home gardens and market gardens, season-extending solar greenhouses, and both home and commercial canning, drying, and freezing. We may consume fewer summer garden vegetables in winter than we do now because using scarce resources to ship fresh produce long distances will become expensive. Probably more of the greens we eat during the off-season will be varieties we can grow under cover—tasty stuff like chard and arugula that isn't too tough to take. Over the course of the whole year we may well eat more produce than we do today, and it will be better produce.

It is highly unlikely that New England will ever produce the bulk of its own grains, oils, meat, and natural textile fibers—many of the basic agricultural commodities of today. We haven't done that for nearly two centuries now, and until our postindustrial population resembles our preindustrial population we needn't even contemplate it. We are going to have to purchase these goods, and if they are sustainably produced they will undoubtedly cost more than they do now. But this higher price will encourage us to grow more grain and livestock products than we presently do, as part of a limited revival of pasture, hay, and grain farming that supports our market gardens.

In particular, we should be able to again supply most of our own dairy products when distant concentrated herds become unprofitable and we finally master rotational pasture management on our home ground. There may be more goats as part of that dairy herd—my crystal ball grows a bit cloudy when it comes to goats. We can surely provide a large part of our own poultry and eggs again, along with some pork, lamb, and wool. These stock fit well into a diversified small farm agroecology that is suited to our region. Such farms, however, will be left a restricted portion of the landscape to work with given our high population on the one hand and the need to protect substantial forest acreage on the other. We will grow what we can, and it will not be enough.

Thus, it is likely that we in New England will import a large part of our grain and meat from the Midwest, as we have since the appearance of canals and railroads. But that's fine. Grain is a very sensible thing to ship, and grassland—the midcontinent analog to our forest, which also needs to be partially restored—is perfect for raising animals. Consuming less meat, but leaner, more flavorful, and healthful meat, is in our best interest. I will cheerfully eat prairie beef or bison, whichever my Kansas ranching friends determine is best for their land. We will doubtless continue to rely on the corn belt for supplemental feedgrain for our dairy, meat, and poultry animals. We will have to pay a price that compensates farmers in the Heartland to provide these commodities sustainably, without ruining their soil in the long run. Through these imports we will bring not only food but also a steady subsidy of crop nutrients into our region—provided we can develop effective, safe systems to compost and recycle organic waste at the household, farm, municipal, and metropolitan scales, instead of defiling and discarding it.

From our wetlands and marginal uplands at the forest's edge will come our regional specialty crops such as cranberries, blueberries, and maple syrup. We can't miss here—we may even go on exporting some of these delicacies from the region as we do now. We can hope to produce more of our own finer beverages in the form of apple cider and fruit wines. We will surely brew more local beer, although I presume we will import most of the barley that goes into it. Indeed, microbrews are on the rise, and after sampling a few I am ready to declare that cider is not far behind. We will certainly continue to import the distinctive crops of other regions of the world: citrus fruits and bananas, coffee, chocolate and tea, olive oil, fine wines, and spices. Such imports may become more expensive if produced and shipped in ways that are ecologically sustainable and socially just, but New Englanders will continue to buy them because they are so good.

Whenever you propose eating closer to home and more in season, some troublemaker is sure to object, "Does this mean I can't have my coffee?" This is where too many overzealous local food system advocates get tongue-tied and choke on their own dry dogma. The wise response is, "Of course you can have your coffee—but why not make sure you buy from a source you can admire?" I don't drink coffee myself, but I have no wish to alienate coffee drinkers (my wife, for example). I like Chilean wines—I just want their apples and grapes to stay home and their campesinos to get a fairer shake. I want New Zealanders to send me their fat, grassy sauvignon blanc and keep their fat, grass-fed lamb. The world of common agrarianism will be made up of distinctive regional cuisines that are largely locally supplied and inspired, together with a flourishing cosmopolitan trade in the delightful crops that are special to each region and desired everywhere. To be purely cosmopolitan, as the globalizers suggest we must, is to be adrift without a cultural compass; to be purely stuck in home mud is to be dully provincial.

Thus far foods; what of wood products? I have suggested that it is our responsibility to keep a good half of our land in natural forest, managed for the protection of biodiversity, recreational enjoyment, and production of wood. Some of that forest should be wild, the bulk of it sustainably used. Given buildings that are well insulated and properly oriented to the sun, wood fuel will be able to provide most of the supplemental space heating for rural and semirural New England towns and perhaps even some electricity in the most forested regions. But heating with wood will remain inappropriate for large urban areas, where energy demand is high and air pollution collects. There, options like cogeneration of electricity and heat by efficient gas turbines make better sense. Wood pulp will still flow from the forest to recharge the stream of recycled paper, but as a by-product of regular thinning for timber improvement rather than from short-rotation clear-cuts. Rural New England towns should be able to supply most of their own construction lumber, but we are hardly going to go back to building our cities primarily with wood. Nevertheless, our architecture and interiors should serve to remind us daily that we live among trees, in a forest whose health and beauty are as indispensable to our own health as they are to our souls. Let us hope that once detached postmodernism has withered and died, strong bioregional architectural styles will spring up, expressing the connection between community and land.

That is a sketch of an attractive, ecologically workable agrarian world of local economies that we would do well to move toward. We need such economies if we are to fully enjoy the world, and we need such economies if we are to survive without fouling the world. Of course, we also need parallel developments (which I have hardly touched on) that make our cities and inner suburbs green and livable—not everyone wants to move to the country, nor should they. The question that follows is, How can we make progress toward that economy and culture in today's world of cheap industrial goods and rampant suburban development? In other words, given a world ruled by largely unrestrained free enterprise and private landownership, which is disinclined to pay much more than lip service to ecological values because that would impose costs that reduce immediate profits, how do we move forward? What do we do with the land once it is protected, if we can't immediately make it pay? One important part of the answer is for private farmers and woodsmen to develop niche markets in the current economy that attract more customers to high-quality local products. Another important part of the answer should be to greatly enlarge the constituency for such local economies by establishing community farms.

What can community farms do that small private farms cannot do better? Teach people about sustainable farming and forestry. I have no doubt that private entrepreneurs will constitute the vertebrae of local economies. Small private farmers will supply most of the production within regional food systems, for the reasons outlined above. The forest may one day be largely owned by conservation commissions and managed by community foresters, but most of the actual logging can probably be hired to private contractors, as it is today. Community farm and forest programs may add something to the production of local natural goods, but that is not their most important contribution. The function of community farms is to help get us from here to there, and to keep us there. Community farms are meant to be educational, even inspirational—to pioneer and perfect sustainable methods before they become economically viable and to involve as many people as possible with the land. Their job is to build popular enthusiasm for protected land and for local products and to grow young farmers and foresters. Although private farmers and loggers will produce the bulk of the output, I believe there is a special place for one community farm and forest organization in every town.

Community farms like Land's Sake are nonprofit entities, incorporated to instruct young people in ecologically sustainable land use. As nonprofits they enjoy such economic advantages as exemption from sales and income taxes and the ability to raise money from tax-deductible donations. While they should pay their workers decent wages and benefits, they are legally bound to devote their funds to educational purposes, not to anyone's profit. They should be designed precisely to work with sustainable, labor-intensive methods, many of which are economically marginal in the present market. I have detailed throughout this book how Land's Sake tries to balance educational goals with economic reality. Community farms operate in today's economy with tomorrow's *ecological* imperatives. What they gain in tax advantages and public support they expend in research and education, by virtue of the sustainable methods they employ. They are ideal grounds for perfecting the use of organic gardens, cider orchards, minor breed livestock, draft horses, agroforestry, and low-impact logging. Not every community farm can afford to try all these ideas, but together they can be a seedbed of better ways of caring for land.

Even more important, community farms involve many people with the land who would otherwise never have that opportunity. There is room for only a handful of full-time farmers today, much as we may hope the number will grow as demand for local, organic food increases. More people can keep their hand in as part-time or hobby farmers—even most commercial farmers today require a second income, and sometimes a third, to survive. Hobby farming helps all farmers by patronizing feed stores, equipment dealers, slaughterhouses, and the like, propping up the agricultural infrastructure of the region. But most would-be farmers do not have the chance to farm their own land, even in a limited way. They may, however, help farm common land.

Many such folks have volunteered at Land's Sake over the years, at the Case Farm, at the sugarhouse, with the livestock, and even in the woods. Often they work in exchange for fresh produce, not to mention the exercise and enjoyment. Sometimes informal partnerships result—for example, for several years a woman helped Land's Sake harvest and arrange dried flowers in return for a supply for her own business. The same sort of exchange might be possible with a woodworker who helps in the forest or with a cider maker who helps care for an orchard. If a person wants to spend a few hours every week working on the land to serve the community, to garner fresh salad greens and pesto, or just to enjoy being outdoors and sweating a little, a community farm gives them that chance. Indeed, that is what I do now. While working one afternoon at the farm I bumped into Mike Potter (for a time our sawyer and farm manager and now a physician), who was out pulling witchgrass in the rhubarb patch. When I crowed about this on the phone to Mike's brother John, who now manages land for the Land Bank on Martha's Vineyard, he laughed and said, "Well, you have that luxury." What he also meant was, you have that obligation. Then he went out to dig clams on some common mudflat he had recently protected.

Most residents become involved with farms in their towns simply through coming to pick flowers, berries, and vegetables for themselves. This may not sound like much, but believe me, it builds support for local agriculture like nothing else. Community farms thrive on pick-your-own sales. They are tailor-made to organize farm festivals and harvest fairs and to lead educational workshops and forest walks—things that tend to be a real headache for hard-driving commercial operators. Community farms are also a perfect means to funnel fresh produce to food banks and shelters, in the manner of Green Power and the Harvest for Hunger. In all these ways, community farms promote the agricultural cause and draw attention to its importance for a healthy community.

I am delighted that several former Land's Sake workers have started their own farms. New farmers have always been our highest and best yield. For all that community farming and forestry programs can do to manage land and to involve adults, far and away their most important role is to teach young people to care for land. Even those who do not become farmers and foresters themselves grow up with a deeper appreciation for the land at their doorsteps. I know it is a cliché, but the value of getting kids' fingers in touch with the soil and with the grain of wood cannot possibly be overstated. This dirt and splinters brand of environmental education makes the difference between seeing nature as a dispensable form of entertainment and feeling it as an essential part of our lives.

Knowing that their children are involved with the land in this way rubs off on parents, too. Like it or not, open space is a side issue to most people; their children and the schools, central. Connecting children with the land is a sure way to bring the value of owning conservation land home to residents of the suburbs. How are we to rally support for protecting great swaths of common land, how are we to get our neighbors concerned about such arcane issues as local food systems and the ecological integrity of forests, if we have nothing to attract their attention? It is essential to have a community farming and forestry program in every town, employing young people and making these visionary ideas immediate and vital.

We will not get far in protecting land and rebuilding local economies until we can awaken people's willingness to swim against the tide. This is the place for community farms: helping people redefine what is valuable. Community farms succeed in an economy that pushes cheap food because people are willing to subsidize them in several crucial ways. But *subsidize* is the wrong word—these farms embody a new set of economic values in our culture, a deep current that is slowly gathering strength. First, the land itself is available because townspeople pay to protect it as open space—presently more for esthetic reasons than out of concern for food, but that is a good place to start. Second, local organic farms thrive because a growing number of eaters are willing to search out and pay a good price for top quality food. Third, community farms work because residents support them through their memberships and donations, primarily because they like what these organizations offer to young people. Finally, community farms succeed because the young people who run them value the work far beyond its rather meager monetary compensation. Community farms provide things that people are finding they like to have in the places where they live. A good relation with the land is something that the marketplace cannot provide, any more than it can make a good marriage or raise a healthy child.

Buying common land and local organic produce, supporting community farms, and devoting part of one's life to such work instead of scrambling to get ahead in a career—all help restore healthy land, community service, and satisfying work to economic calculus. Agriculture and forest industries should *not* be driven to the bottom line of extracting food and lumber in the most "efficient," cheapest way. The care of land and community must rise above the bottom line. That is the fundamental, revolutionary principle that community farms express so well.

Only a small minority of Americans will make their living primarily from the land for the foreseeable future, even in rural areas. If we are to build a society capable of taking good care of the land and living sustainably, however, it is vitally important that most of us have some meaningful connection with the land. The restless drive of many Americans to return to rural places presents the opportunity and indeed the necessity of building a new, common agrarianism in the suburbs. The suburbs are something we have to deal with. If we can manage a working balance of residential development, farmland, and forests as the suburbs spread, we have the chance to involve people with the land in many small but telling ways. We can make living with the land an integral part of most American communities again—not something left entirely to a few stressed farming professionals, but part of the daily (or at least weekly) lives of a large part of the population.

There is a place for common land and community farms right across this country. The idea may be most obviously suitable to New England, where community identity and the habit of local democratic control are well entrenched, but I am sure it is adaptable to other regions. In places where rural land is not incorporated into towns as it is here, locally controlled land trusts can stand in for town government when it comes to acquiring common land. In sparsely populated areas in the nation's midsection, it might be best to work at the county level. The proper place for common land is something that each region can determine for itself, according to its own geographic and political character.

Needless to say, just what land use takes place on the commons will vary greatly as well. Immediately surrounding the cities, dozens of community market gardens like our Case Farm can dramatically alter the face of suburbia, staving off the finality of build-out. Community farms can supply an important source of locally grown food in the process. A bit farther out, if more common land is protected, community farms can add mixed husbandry and agroforestry projects to their agenda and play a powerful educational role in keeping local agriculture alive. In the hinterlands, I can envision a great common forest composed of thousands of connected local common forests, stretching from Arkansas to Maine along the mountain chain, back to Minnesota through the North Woods, and merging into the common prairies beyond. The purpose of this reclaimed commons must be not only to protect ecologically essential land for the common good, but also to protect the livelihoods of the people inhabiting threatened rural communities, and to bring newcomers to a more meaningful, useful sense of the places to which they are moving.

Every place is important. We need not search for some postcard village, with its face timelessly pastoral and its back to pristine wilderness. That place no longer exists, if it ever did. If it were found today, it would be quickly consumed by its discoverers. We need instead to dig in wherever we are. If we are to build a sustainable society, we must work with people where they live now. We must change minds about how land is to be used and about what values are to be accounted by our economy. This means guarding land while growing food, cutting trees while preserving biodiversity, and generally coming to grips with the suburbs in all their messiness and contradiction.

We must recover the traditions that once shaped places in distinctive ways, traditions reaching back in our peasant memory to dreams of both secure private ownership and access to commons, before the shopping malls and tract houses obliterate them. We must both honor and excel our rural forebears, who did not see clearly enough that excessive individualism in a market economy was the outstretched neck of agrarianism, unwary of the ax. We must rebuild functioning communities with closer ties to the land not just in nostalgic fantasy, not just in token preservation, but in substantial daily practice. We must reclaim the commons.

Reclaiming the commons. (Susan M. Campbell)

Appendix

An Approach to Local Engagement

The thinkers who have inspired me are obvious enough: Henry Thoreau, Aldo Leopold, Wendell Berry, Donald Worster, Wes Jackson. The reader will find their work (and that of other good thinkers) in the notes to this book. That work will lead on to other work, both through the notes and through neighboring volumes on the same shelf in the stacks, which is how I have discovered many of the writers I treasure the most. Most of my other references are to primary and secondary sources peculiar to my part of the world. What I offer here is not so much a guide to further reading as a general approach to knowing and caring for particular places. This approach can be pursued by an individual or adapted by a skillful teacher to a school course from the elementary to graduate level.

Knowing

To know a place, first get out in all weathers. Walk the land at least; at best, work the land. In this way the subtleties of place become familiar. But beyond direct experience, there are documentary aids that can help structure understanding. Start by pinning the local U.S. Geological Survey (USGS) topographic quadrangle to your wall.

Think of the place where you live as made up of layers—geological, biological, cultural. These layers are not simply piled one on top of the next but exist simultaneously and converse with one another. All of them are changing, but at varying rates. The conceptual key to placing yourself in time is to remember this: we live not at the *end* of these often cyclical processes (this is the self-congratulatory Progressive view), but in the *midst* of them.

Geology is fundamental to the structure of the land—its topography, soils, flow of water, zones of vegetation. In Massachusetts, the surficial geology—the variety of ways land was deposited by the glacier—is crucial. The USGS has colorful Surficial Geology maps for many quadrangles. In Kansas, I discovered, variations in bedrock geology were what counted most—the way layers of sandstone, limestone, and shale were deposited and subsequently dissected by water. Most states also have geological surveys (often associated with a state university), which can supply maps and point you to the regional texts that will help make sense of those maps.

Soils are made from the parent materials supplied by crumbled rock. Very different soils are often found in different parts of the local landscape. In my area, soils formed in glacial till are a long way in character from those formed in outwash, which are far from the mucks and peats of low-lying areas. The Natural Resources Conservation Service prepares Soil Surveys for every county. These detailed soil maps, with accompanying texts explaining how the soils have been classified, grouped, and assessed as to their capabilities for agriculture and other uses, are indispensable to knowing a place.

Upon the soil grows vegetation, which in turn helps generate soil from stone. Every landscape has a characteristic pattern of ecosystems upon it. The Audubon Society, Sierra Club, and similar groups often publish first-rate field guides and natural histories to these ecosystems. The notes in the better guides lead to sources that are more technical, for further study. Remember that although ecosystems correspond in a broad way to underlying topography and soils, the resulting patterns are not fixed. They change according to natural disturbances, changing climate, species migration, and human influences. To follow the shifting of your mosaic you must venture into the paleoecological literature. Proceedings of symposia and the like appear every few years, reviewing current knowledge of past climates and vegetation around the world. These will help you get situated and lead you to more specific studies concerning your own region.

Next is the history and prehistory of human inhabitants. Most regions have accumulated an anthropological literature about how native peoples lived—of course, you might also talk to native people themselves, for another perspective. Crossing into historical European (and, in some cases, African or Asian) inhabitancy, you will discover a growing literature in environmental, agricultural, and social history. You may not find a detailed study of your home place, but among these disciplines you will likely discover several about similar places not too far away.

To investigate your own place (either your community or a particular neighborhood or piece of land within it), you will need to get into primary materials. Often you will find a privately published local history, but unless you are very lucky the author will skip over boring little details like how people actually lived on the land and get right to the famous battle that once took place in your neck of the woods. You may find collections of anecdotes and reminiscences, but it will be up to you to build an ecological framework. You must get beyond the "in the old days life was tough, but people worked harder, grew everything they needed for themselves, and looked out for each other more than we do today" clichés about the pioneer era—which, in the popular mind, was anytime before World War II. It will at least be instructive to discover that people 150 years ago wrote the same things about *their* grandparents.

It is possible to map the history of your place. This is what I do. Over much of the nation's heartland you can start from the nineteenth-century public land survey—you may be able to find the original survey notes, too. Elsewhere (such as in New England) you can begin with the earliest records of land divisions, two or three centuries ago. The chain of landownership can then be followed through deeds and probated estates down to the present and mapped. The way in which farm properties were arranged upon the face of the land—over topography and soil—is revealing. Reconstructing these patterns is painstaking, tedious work but very rewarding in my experience. It is possible to do it on paper or acetate overlays, but I strongly recommend some kind of Geographical Information Systems (GIS) mapping software. These can now be run on a reasonably powerful home computer—a 486 or Pentium or Macintosh equivalent. I use a program called MapInfo on a Power Mac.

The object of mapping landownership is to get at changing patterns of land use. What local resources were people using, where were they finding them, how did farming and logging change the face of the land? There actually are some data on what people grew. The U.S. Agricultural Census has collected this information about once a decade from 1850 to the present. The published data are aggregated by state and by county and are readily available. You can find agricultural returns for individuals like your great-grandfather from 1850 to 1890 on microfilm at your regional Federal Records Office. Many states conducted an agricultural census of their own, especially in the late nineteenth and early twentieth centuries. These data were sometimes aggregated at the township level. In some places you may also be able to glean land use information from property tax records. For example, the tax valuation information from Massachusetts towns and farmers between 1749 and 1860 is just phenomenal. Many dissertations, including mine, have been based largely on these data.

Finally, once you have looked at the rocks and the trees and the history, there is one more layer to consult: the art. Every place has its literature. Become a close, critical student of regional writers. Visual artists have also interpreted your place and continue to engage with it. Get to know this body of work and these people—photographers, craftspeople, painters, and poets. Somewhere near you is a man who weaves fantastic designs out of native shrubbery, a woman who paints walls with multicolored mud. Some you will not choose to accompany very far on their journey, but others will open your eyes in new ways.

Studying a place obliges anyone to venture into unfamiliar fields of scholarship and to gain a working understanding of what goes on there. Often this involves a daunting new vocabulary and a long history of interpretive quarrels. I am assuming that most of my readers are college graduates or in any case know their way around a library. I am also assuming that few, even among environmentalists, have explored all the disciplines that might bear on knowing one place. I often find myself far afield. As Wes Jackson puts it, our universities offer many majors in upward mobility but none in homecoming.

Here is the seat-of-the-pants way I break into a new field, given a decent university library. Usually I begin with one or two references that sound vaguely interesting. These get me to the right part of the stacks, where I can browse through other interesting titles. I find the good gray texts, the sturdy monographs, and go straight to the footnotes for more references. Soon I discover the respected scholars and journals in the field and the crucial review articles that appear from time to time. Next, I search for more works by those authors; go browse the recent issues of those journals. Somewhere along the way I stumble upon a few studies that are most useful to me and meanwhile gain some understanding of their scholarly context. Each field has its own way of defining issues, its own way of evaluating sources, its own way of talking with which you must become conversant, without becoming hopelessly befuddled.

The other thing you must do is cultivate people. You can talk to eminent authorities in academia and state agencies—of course you will get a wide range of responses but most will be pleased to help you out and point you to more good sources. That is their business, after all. But also cultivate local teachers—birdwatchers, mushroom hunters, flint collectors, and antiquarians; individually and in their amiable societies. These people know their way around. Many of them are enjoyably eccentric, as they tend to be independent thinkers and free spirits. You may have to put what they tell you through a fine mesh, but in the meantime, your sense of place will be immeasurably enriched.

All of what I have laid out here about knowing your place can be undertaken by a lone individual, following his or her own bent. Indeed that is how the best, idiosyncratic work is often done. But local studies done in this way can also form the basis of a college course. Knowing our place can and should also be built into the curricula of public schools, at every level. This is the way I have been promoting it with teachers in recent years, and I heartily recommend it to educators everywhere.

Caring

Since this entire book presents one model of community farming and forestry, I do not need to repeat myself here, except to add a few observations. One is that the free spirits mentioned above, who will help you get to know and love a place, are often not the folks who can best help you accomplish anything there. You must learn who is taken seriously in the community and make sure that such people are identified with your organization, without losing the creative energy of the free thinkers.

Along similar lines, those who come into a community with a preconceived agenda and go charging off before they know the lay of the land will find themselves perpetually perplexed that no one is taking their sensible ideas seriously. It took me about five years to figure out more or less how Weston worked. Find out who listens to whom.

It is very helpful to make contact with the wider community of people and organizations involved in similar efforts. You will want to know about other organic farmers, sustainable foresters, land trusts, and bioregionalists. A few years ago, after the publication of Wendell Berry's *Another Turn of the Crank,* the Land Institute prepared a brief Resource Guide of umbrella organizations and networks along these lines. I suggest you call the Land Institute at 785-823-5376 or write them at 2440 East Water Well Road, Salina, Kansas, 67401, to request this guide. While you're at it, become a Friend of the Land and receive the *Land Report,* which will keep you abreast of some of the most innovative research in sustainable agriculture going on today.

Finally, let me repeat that the most important quality of community farming is tenacity. Devotion to place means just that. Do not give up.

Notes

Introduction: Wilderness and Suburbia

1. Roderick Nash, *Wilderness and the American Mind,* 3d ed.(New Haven: Yale University Press, 1982); Samuel P. Hays, *Beauty, Health, and Permanence: Environmental Politics in the United States, 1955–1985,* in collaboration with Barbara D. Hays (New York: Cambridge University Press, 1987).
2. For example, Michael Pollan, *Second Nature: A Gardener's Education* (New York: Atlantic Monthly Press, 1991); William Cronon, "The Trouble with Wilderness; or, Getting Back to the Wrong Nature," in William Cronon, ed., *Uncommon Ground: Toward Reinventing Nature* (New York: Norton, 1995).

Chapter 1. Green Power and Land's Sake

1. Berry quotes the Sioux holy man in *A Continuous Harmony: Essays Cultural and Agricultural* (New York: Harcourt, Brace Jovanovich, 1975), 126–27.
2. Nancy M. Fleming, *Weston Town Commons: A History* (Weston, Mass.: Weston Garden Club, 1991). See also John R. Stilgoe, "Town Common and Village Green in New England: 1620 to 1981," in Ronald Lee Fleming and Lauri A.

Halderman, ed., *On Common Ground: Caring for Shared Land from Town Common to Urban Park* (Harvard, Mass.: Harvard Common Press, 1982).

3. A term coined by Garret Hardin. See chap. 6 for a full discussion.
4. See Robert McCullough, *The Landscape of Community: A History of Communal Forests in New England* (Hanover, N.H.: University Press of New England, 1995), for an excellent overview of the protection of common land in New England.
5. William A. Elliston, "Oral History of the Town of Weston," interviewed by Jeanette Bailey Cheek, 1981, typescript in Weston Public Library.
6. I quote Hanson's speech the way I remember it, so the wording may not be strictly accurate.

Chapter 2. Market Gardening

1. Peter L. Engel and the Cornucopia Project of Rodale Press, *The Massachusetts Food System: Leading the Nation in Vulnerability or Forging a Path to Self-Reliance?* (Emmaus, Penn.: Rodale Press, 1983).
2. My figures are drawn from various U.S. and Massachusetts agricultural censuses. Historians will be aware that census figures are fraught with difficulties because of incomplete reporting and changes in classification. These data should therefore be regarded as revealing broad trends, not as precisely accurate.
3. I discuss nineteenth-century Massachusetts agriculture in "Skinning the Land: Economic Growth and the Ecology of Farming in Nineteenth-Century Massachusetts," paper presented at the American Social History Society, Chicago, November 1988. For a similar analysis, see Michael M. Bell, "Did New England Go Downhill?" *Geographical Review* 79 (1989). A good overview of New England agriculture is Howard S. Russell, *A Long, Deep Furrow: Three Centuries of Farming in New England* (Hanover, N.H.: University Press of New England, 1976). Classic reference works for the nation as a whole, and still good, include Paul Wallace Gates, *The Farmers' Age: Agriculture 1815–1860* (New York: Holt, Rinehart and Winston, 1960); Fred A. Shannon, *The Farmers' Last Frontier: Agriculture, 1860–1897* (New York: Farrar and Rinehart, 1945); and Harold U. Faulkner, *The Decline of Laissez-Faire, 1897–1917* (New York: Rinehart, 1951).
4. Faulkner, *The Decline of Laissez-Faire*, 335.
5. Richard A. Wines, *Fertilizer in America: From Waste Recycling to Resource Exploitation* (Philadelphia: Temple University Press, 1985); James Wharton, *Before Silent Spring: Pesticides and Public Health in Pre-DDT America* (Princeton: Princeton University Press, 1974).
6. A. H. Smith, "Market Gardening," *Agriculture of Massachusetts 1888* (Boston: Board of Agriculture, 1889); R. L. Watts, "Recent Advancements in Market Gardening," *Agriculture of Massachusetts 1912* (Boston: Board of Agriculture, 1913).

7. Two important books on water in the West are Marc P. Reisner, *Cadillac Desert: The American West and Its Disappearing Water* (New York: Viking, 1986); and Donald Worster, *Rivers of Empire: Water, Aridity, and the Growth of the American West* (New York: Pantheon, 1985).
8. The following discussion is not meant to be a thorough, point-by-point conviction of industrial agriculture but merely a summary of the main points of the indictment. For more complete discussions, see Judith D. Soule and Jon K. Piper, *Farming in Nature's Image* (Washington, D.C.: Island Press, 1992); Wes Jackson, Wendell Berry, and Bruce Coleman, eds., *Meeting the Expectations of the Land* (San Francisco: North Point Press, 1984). For an opposing ecological endorsement of industrial farming, see, for example, Dennis T. Avery, *Biodiversity: Saving Species with Biotechnology* (Indianapolis: Hudson Institute, 1993).
9. See Sandra Postel, *Dividing the Waters: Food Security, Ecosystem Health, and the New Politics of Scarcity* (Washington, D.C.: Worldwatch Institute, 1996), for a review of aquifer and surface water depletion. Worldwatch Papers and annual *State of the World* compilations give a good overview of many of the concerns sketched here.
10. Janet N. Abramovitz, *Imperiled Waters, Impoverished Future: The Decline of Freshwater Ecosystems* (Washington, D.C.: Worldwatch Institute, 1996), addresses many of these concerns.
11. Angus Wright, *The Death of Ramón González: The Modern Agricultural Dilemma* (Austin: University of Texas Press, 1990) is a look at the health impact of modern food systems on Third World farmworkers. Two recent nonpolemical books on the continuing ineffective regulation and unnecessary overuse of pesticides are John Wargo, *Our Children's Toxic Legacy: How Science and Law Fail to Protect Us from Pesticides* (New Haven: Yale University Press, 1996); and Mark L. Winston, *Nature Wars: People vs. Pests* (Cambridge: Harvard University Press, 1997).

Chapter 3. Livestock and Grass

1. U.S. Census of Agriculture, 1850, 1978, 1987.
2. Russell, *A Long, Deep Furrow,* was the first to bring this pattern to my attention.
3. *Watertown Records: First and Second Books of Town Proceedings* (Watertown, Mass.: Press of Fred G. Barker, 1894), 1:8, 21–24, 52, 94–99, 104–05, 111, 146–147; 2:3–5.
4. In 1761 a group of meadow owners petitioned the General Court of Massachusetts to form a Commission of Sewers to drain the meadow near Weston Center. A Commission of Sewers was a venerable English institution whose purpose was to provide for common management of a waterway to improve adjoining property. Massachusetts Archives 1:366.

5. An example of the right to turn water on meadow in Weston can be found in a deed of James Jones to his son Isaac, Middlesex Registry of Deeds book 59, page 570, 1762. A good discussion of the idea of a competency is Daniel Vickers, "Competency and Competition: Economic Culture in Early America," *William and Mary Quarterly* 47 (1990).
6. A much more detailed analysis of colonial husbandry in the neighboring town of Concord can be found in my dissertation, "Plowland, Pastureland, Woodland and Meadow: Husbandry in Concord, Massachusetts, 1635–1771," Brandeis University (1995).
7. For a splendid story of women's work in New England, see Laura Thatcher Uhlrich, *The Midwife's Tale: The Life of Martha Ballard, Based on Her Diary, 1785–1812* (New York: Knopf, 1990).
8. There are many fine works on the process by which New England towns built up demographic pressure. See Philip J. Greven, Jr., *Four Generations: Population, Land, and Family in Colonial Andover, Massachusetts* (Ithaca: Cornell University Press, 1970); Robert A. Gross, *The Minutemen and Their World* (New York: Hill and Wang, 1976); Kenneth A. Lockridge, *A New England Town: The First Hundred Years* (New York: Norton, 1970); Carolyn Merchant, *Ecological Revolutions: Nature, Gender and Science in New England* (Chapel Hill: University of North Carolina Press, 1989).
9. J. M. Smith, Franklin County Agricultural Survey, *Agriculture of Massachusetts, 1865* (Boston: Board of Agriculture 1866), 307. The following discussion is more fully documented in my "Skinning the Land."
10. Francis DeWitt, *Statistical Information of Branches of Industry in Massachusetts, 1855* (Boston: William White, 1856); Horace C. Wadlin, "Fisheries, Commerce, and Agriculture," *Census of Massachusetts 1895* (Boston: Wright and Potter, 1899).
11. Good discussions of western grain farming can be found in Shannon, *Farmers' Last Frontier;* and William Cronon, *Nature's Metropolis: Chicago and the Great West* (New York: W. W. Norton, 1990).
12. The classic work here is Donald Worster, *Dust Bowl: The Southern Plains in the 1930s* (New York: Oxford University Press, 1979).
13. Thanks to Ellen and Roy Raja for a discussion of their enterprise, and for the emergency phone calls over the years.
14. One of the most concise statements of the importance of beauty in our everyday surroundings that I know is Gregory Conniff, "Where do you love?" *The Land Report* 50 (1994).
15. Vance Nye Bourjaily, *Country Matters* (New York: Dial Press, 1973).

Chapter 4. Tree Crops

1. Howard S. Russell, *Indian New England Before the Mayflower* (Hanover, N.H.: University of New Hampshire Press, 1980). Marjorie Green Winkler, "A 12,000-Year History of Vegetation and Climate for Cape Cod, Massachusetts," *Quaternary Research* 23 (1985).
2. Weston Town Report, 1894, Weston Public Library.
3. David R. Houston, Douglas C. Allen, and Denis Lachance, *Sugarbush Management: A Guide to Maintaining Tree Health* (Radnor, Penn.: USDA Northeastern Forest Experiment Station, 1990).
4. Russell, *Long, Deep Furrow.*
5. Sarah F. MacMahon, "A Comfortable Subsistence: The Changing Composition of Diet in Rural New England, 1620–1840," *William and Mary Quarterly* 42 (1985), details the change from beer to cider.
6. Henry David Thoreau, "Wild Apples," *The Natural History Essays* (Salt Lake City: Peregrine Smith, 1980), 209; Edward Jarvis, *Traditions and Reminiscences of Concord, Massachusetts, 1779–1878* (Amherst, Mass.: University of Massachusetts Press, 1993); William J. Rorabaugh, *The Alcoholic Republic: An America Tradition* (New York: Oxford University Press, 1979).
7. Wharton, *Before Silent Spring;* S. T. Maynard, "Insecticides and Fungicides, and their Practical Application," *Agriculture of Massachusetts, 1894* (Boston: Board of Agriculture, 1894); A. H. Kirkland, "Insecticides," *Agriculture of Massachusetts, 1895* (Boston: Board of Agrculture, 1896); *Agriculture of Massachusetts, 1899* (Boston: Board of Agriculture, 1900), 104.
8. I. A. Merwin et al., "Scab-resistant apples for the Northeastern USA: New prospects and old problems," *Plant Disease* 78 (1994). Thanks to Dan Cooley and Bill Coli for bringing the idea of a scab-tolerant land race to my attention.
9. *The Journal of Henry D. Thoreau* (Boston: Houghton Mifflin, 1906), 5:509–10; Brenton H. Dickson, *Once Upon a Pung* (Boston: Thomas Todd Company, 1963).
10. See my "'Dammed at Both Ends and Cursed in the Middle': The 'Flowage' of the Concord River Meadows, 1798–1862," *Environmental Review* 13 (1989). Cranberry statistics from Massachusetts Agricultural Census.
11. K. H. Deubert and F. L. Caruso, "Bogs and Cranberry Bogs in Southeastern Massachusetts," University of Massachusetts Cranberry Experiment Station, Research Bulletin 727 (1989).
12. Thoreau, "Huckleberries," *Natural History Essays,* 227–28.
13. Ibid., 245; Ralph Waldo Emerson, "Thoreau," *The Complete Works of Ralph Waldo Emerson,* ed. Edward Waldo Emerson (Boston: Houghton, Mifflin, 1904), vol. 10.

14. Alice Tyler Fraser, "Growing up in Weston, 1903–1920," *Weston Historical Society Bulletin* 21 (1984).
15. United States Agricultural Census; Thoreau, "Huckleberries," 249.
16. For a good discussion of coppicing and woodlands, see Oliver Rackham, *The History of the Countryside* (London: J. M. Dent, 1986).
17. Russell, *Indian New England;* Peter Thomas, "Contrasting Subsistence Strategies and Land Use as Factors for Understanding Indian-White Relations in New England," *Ethnohistory* 23 (1976). The 20 percent estimate is from Mitchell T. Mulholland, "Territoriality and Horticulture: A Perspective for Prehistoric Southern New England," in George P. Nicholas, ed., *Holocene Human Ecology in Northeastern North America* (New York: Plenum Press, 1988). Mulholland argues that by the time of European contact, the natives of southern New England may have pressed their arable land to the limit. Also see discussions in Cronon, *Changes in the Land,* and Merchant, *Ecological Revolutions.*
18. The only evidence of meadow burning I am aware of is circumstantial: there were many open grassy meadows when the English arrived, and these have become forested swamps today.
19. My favorite review of the Indian burning debate is William A. Patterson and Kenneth E. Sassaman, "Indian Fires in the Prehistory of New England," Nicholas, ed., *Holocene Human Ecology in Northeastern North America.* Hirtleberry, or whortleberry, is an early English name for blueberry.
20. Two very accessible reviews include Daniel Hillel, *Out of the Earth: Civilization and the Life of the Soil* (New York: Free Press, 1990); and Otto T. Solbrig and Dorothy J. Solbrig, *So Shall You Reap: Farming and Crops in Human Affairs* (Washington, D.C.: Island Press, 1994).
21. See Wes Jackson, *New Roots for Agriculture* (San Francisco: Friends of the Earth, 1980), for the original statement of the case for herbaceous seed-bearing perennial polycultures. Soule and Piper, *Farming in Nature's Image,* provides a more recent synthesis.

Chapter 5. The Town Forest

1. William S. Broecker and George H. Denton, "What Drives Glacial Cycles?" *Scientific American* 262 (1990), is an accessible overview of glacial cycles and forcing mechanisms. Steven M. Stanley and William F. Ruddiman, "Neogene Ice Age in the North Atlantic Region: Climatic Changes, Biotic Effects, and Forcing Factors," *Effects of Past Global Change on Life* (Washington, D.C.: National Academy Press, 1995), provides a more recent, technical review.
2. Thompson Webb III, "Pollen Records of Late Quaternary Change: Plant Community Rearrangement and Evolutionary Implications," *Effects of Past Global*

Change on Life, is a good introduction to current paleoecological thinking. Margaret B. Davis, "Quaternary History of Deciduous Forests of Eastern North America and Europe," *Annals of the Missouri Botanical Garden* 70 (1983), is one of the classic articles in the field.

3. The "shifting ecological mosaic" was introduced in F. Herbert Bormann and Gene E. Likens, *Pattern and Process in a Forested Ecosystem: Disturbance, Development and the Steady State Based on the Hubbard Brook Ecosystem Study* (New York: Springer-Verlag, 1979). Thompson Webb argues that such shifting never reaches a steady state: see "The Past 11,000 Years of Vegetational Change in Eastern North America," *Bioscience* 31 (1981). Also see Douglas G. Sprugel, "Disturbance, Equilibrium, and Environmental Variability: What Is 'Natural' Vegetation in a Changing Environment?" *Biological Conservation* 58 (1991), for a good discussion of this issue.
4. I am obviously much in sympathy with Daniel Botkin, *Discordant Harmonies: A New Ecology for the Twenty-first Century* (New York: Oxford University Press, 1990), but he seems to draw a different conclusion. At times he prescribes a level of human omniscience and managerial control over nature with which I am not comfortable.
5. For an ecological history of England, see Rackham, *The History of the Countryside.*
6. The essays in Brooke Hindle, ed., *America's Wooden Age: Aspects of Its Early Technology* (Tarrytown, N.Y.: Sleepy Hollow Press, 1975); and id., *Material Culture of the Wooden Age* (Tarrytown, N.Y.: Sleepy Hollow Press, 1981), are good starting points for gaining an understanding of this era.
7. Robert Tarule, "The Joined Furniture of William Searle and Thomas Dennis: A Shop-based Inquiry into the Woodworking Technology of the Seventeenth Century" (Ph.D. diss., Graduate School of the Union Institute, 1992), is my authority here.
8. Charles F. Carroll, *The Timber Economy of Puritan New England* (Providence: Brown University Press, 1973).
9. Tarule; *Watertown Records* 1:2.
10. For a detailed account of the impact of deforestation on rivers, see my "Dammed at Both Ends and Cursed in the Middle."
11. George B. Emerson, *Report on Trees and Shrubs of Massachusetts* (Boston: Little, Brown, 1846).
12. See McCullough, *The Landscape of Community,* for an excellent account of the forestry movement.
13. Michael Pollan, *Second Nature,* tells the story of the political mayhem that follows when such stands fall in isolation from the softening context of a larger

protected forest and the larger understanding that goes with protecting such a forest.

14. George B. Emerson, "Forest Trees," *Transactions of the Norfolk Agricultural Society* (1859).
15. Malcolm Hunter et al., "Paleoecology and the Coarse Filter Approach to Maintaining Biological Diversity," *Conservation Biology* 2 (1988).
16. A fine guide to protecting and managing forest in Massachusetts is Christina M. Petersen et al., *The Forest Use Manual* (Amherst, Mass.: University of Massachusetts Cooperative Extension System, 1992).

Chapter 6. Reclaiming the Commons

1. A few titles include Marty Strange, *Family Farming: A New Economic Vision* (Lincoln, Neb.: University of Nebraska Press, 1988); Gene Logsdon, *At Nature's Pace: Farming and the American Dream* (New York: Pantheon, 1994); id., *The Contrary Farmer* (Post Mills, Vt.: Chelsea Green, 1993); Wendell Berry, *The Unsettling of America: Culture and Agriculture* (San Francisco: Sierra Club, 1977); id., *Another Turn of the Crank* (Washington, D.C.: Counterpoint Press, 1995); Wes Jackson, *Becoming Native to this Place* (Lexington: University Press of Kentucky, 1994).
2. See Gary Snyder, "Good, Wild and Sacred," *The Practice of the Wild* (San Francisco: North Point, 1990), for a discussion along similar lines.
3. Thoreau, "Huckleberries," 255–56.
4. Randall Arendt, *Rural by Design: Maintaining Small Town Character* (Chicago: Planners Press, 1994).
5. Garret Hardin, "The Tragedy of the Commons," *Science* 162 (1968). A good review of commons systems is Bonnie J. McCay and James M. Acheson, eds., *The Question of the Commons: The Cultural Ecology of Communal Resources* (Tucson: University of Arizona Press, 1987).
6. Many of the economic ideas touched upon in this book are argued at length in Herman E. Daly and John B. Cobb, Jr., *For the Common Good: Redirecting the Economy Toward Community, the Environment, and a Sustainable Future* (Boston: Beacon Press, 1994).

Index